MEDIA AND COMMUNICATIONS - TECHNOLOGIES, POLICIES AND CHALLENGES

COGNITIVE RADIO: TERMINOLOGY, TECHNOLOGY AND TECHNIQUES

MEDIA AND COMMUNICATIONS - TECHNOLOGIES, POLICIES AND CHALLENGES

Additional books in this series can be found on Nova's website at:

https://www.novapublishers.com/catalog/index.php?cPath=23_29&seriesp=
Media+and+Communications+-+Technologies%2C+Policies+and+Challenges

Additional e-books in this series can be found on Nova's website at:

https://www.novapublishers.com/catalog/index.php?cPath=23_29&seriespe=
Media+and+Communications+-+Technologies%2C+Policies+and+Challenges

MEDIA AND COMMUNICATIONS - TECHNOLOGIES, POLICIES AND CHALLENGES

COGNITIVE RADIO: TERMINOLOGY, TECHNOLOGY AND TECHNIQUES

PEYTON D. MCGUIRE
AND
HARRY M. ESTRADA
EDITORS

Nova Science Publishers, Inc.
New York

For permission to use material from this book please contact us:
Telephone 631-231-7269; Fax 631-231-8175
Web Site: http://www.novapublishers.com

LIBRARY OF CONGRESS CATALOGING-IN-PUBLICATION DATA

Cognitive radio : terminology, technology, and techniques / editors, Peyton D. McGuire and Harry M. Estrada.
p. cm.
Includes index.
ISBN 978-1-60876-604-8 (hardcover)
1. Cognitive radio networks. I. McGuire, Peyton D. II. Estrada, Harry M.
TK5103.4815.C628 2009
621.384--dc22

2009048223

Published by Nova Science Publishers, Inc. ✢ New York

CONTENTS

PREFACE

Cognitive radio is a radio that can sense, learn and adapt to the surrounding environment according to its inner and outer stimuli. A primary feature of cognitive radios is the ability to adapt the transmission parameters given a dynamic wireless environment. In this book, the cognitive radio adaptation is formalized and four evolutionary algorithms are explored and used to intelligently solve this problem and determine the optimal parameters for a given situation. Moreover, the authors of this book limit the scope of cognition to reduce mutual interference between CR-based rental (unlicensed) users and licensed users (LUs) and in providing coexistence between them. The modulation strategies employed to realize a co-existence between the CR-based rental system and the licensed system are introduced. In addition, a novel dynamic spectrum sharing scheme combining spectrum adaptation and MIMO-OFDM for cognitive radio system is proposed. In order to improve spectrum efficiency and reduce computational complexity, a simple power allocation algorithm is proposed as well. Other chapters report recent results on policy-based self-management towards self-adaptive and Cognitive Radio Systems (CRS). Its focus is not on self-properties of cognitive radio systems but on self-learning policy and cognition.

In Chapter 1 the authors survey the three primary approaches in cognitive radio; namely the interweave, overlay, and underlay approaches. The authors then focus on underlay secondary networks which have several advantages over the other two. For example, underlay networks avoid the error-prone sensing problem and the need for dynamic re-allocation of spectrum associated with the interweave approach. They also do not require the large amount of side information about the primary users needed in the overlay approach. However, conventional underlay networks have to adopt a conservative approach in transmit power levels and, therefore, interference suppression is needed. After an overview of interference mitigation in underlay networks for ultra-wide band (UWB) systems the authors present a novel interference-minimizing code assignment (IMCA) scheme for code-division multiple-access (CDMA) secondary networks. While technically an underlay approach the IMCA scheme opportunistically assigns the secondary user codes available in the code subspace to minimize the mean square crosscorrelations between codes of existing and arriving users. Furthermore, the scheme does not require knowledge of the codes or symbol epoch of the primary CDMA network. It thus attains the advantages of opportunistic and underlay networks and avoids the large side information requirements of overlay systems. Significant performance gains are achieved over random code assignment. The authors then briefly discuss relaying as another technique for extending the range of underlay networks. Finally, the authors present their conclusions and assessment of underlay networks.

A primary feature of cognitive radios is the ability to adapt the transmission parameters given a dynamic wireless environment. In Chapter 2, the cognitive radio adaptation is formalized as a multi-objective optimization problem and four evolutionary algorithms, i.e., genetic algorithms (GAs), quantum genetic algorithm (QGA), particle swarm optimization (PSO), and shuffled frog leaping algorithm (SFLA), are explored and used to intelligently solve this problem and determine the optimal parameters for a given situation.

GAs have been regarded as an efficient algorithm for cognitive radio adaptation. However, classic GAs are quite poor at hill climbing and they may converge to some local maximum. In order to improve the convergence performance of the adaptation process, the authors have proposed to use QGA and PSO for the adaptation method in their previous work. This chapter reviews these three methods and compares their performance. After overview of GAs, QGA and PSO, the authors focus on a newly developed evolutionary algorithm, SFLA, which has not been investigated as thoroughly as GAs or PSO. The authors propose a binary SFLA in this chapter which can be used to solve any combinatorial optimization problem in general, and the proposed binary SFLA is used as the adaptation method for cognitive radios. After investigation binary SFLA-based cognitive radio adaptation, the authors present some challenges and future directions.

The growing demand on wireless communication systems to provide high data rates has brought with it the need for a flexible and efficient use of the spectrum resource, which is a scarce commodity. The regional spectrum allocation policy counteracts the free mobility of radio communication equipment. The vast majority of the available spectral resources have already been licensed, so it appears that there is little or no room to add any new services, unless some of the existing licenses are discontinued. Furthermore, recent studies and measurements have shown that vast portions of the licensed spectra are rarely used due to the inflexible spectrum regulations. The whole idea behind Cognitive Radio (CR) use is that it should prompt effective spectrum use, since intelligence and learning processes aid the radio system to access the spectrum effectively. The CR system has learning and understanding capabilities so that the stated goals may be achieved. In Chapter 3 the authors shall limit the scope of cognition to reduce mutual interference between CR-based rental (unlicensed) users and licensed users (LUs) and in providing coexistence between them. It is expected that the rental users will be allowed to transmit and receive data over portions of spectra when primary (i.e., licensed) users are inactive. In this article the authors will introduce the modulation strategies employed to realize a co-existence between the CR-based rental system and the licensed system. This is done in such a way that the rental users (RUs) are invisible to the licensed users (LUs). The authors consider the rental user accesses the unoccupied LU band in overlay fashion.

Studies of white-space cognitive radio systems are being actively pursued in response to the frequency depletion problem. At present, a signal of lumped frequency is placed in one of the white-space frequency bands, such as TV bands. In the future when cognitive radio techniques are mature, many frequency bands with discrete center frequencies may be used simultaneously in order to transmit extremely broadband data. However, many problems have to be solved in order to realize the system. In view of this prospect, the authors have proposed a broadband short-range white-space cognitive system COGSIP (COGnitive System of Intelligent Processing) to substantiate the problems and investigate them. COGSIP is a personal area network (PAN) system with a data rate of a few Gbps in the microwave band. A COGSIP terminal searches white spaces from the broad microwave band and utilizes many white spaces simultaneously avoiding primary transmissions to realize a data rate of a few Gbps. In Chapter 4, the authors introduce the outline of COGSIP and four related techniques the authors have investigated: high-sensitivity energy detection, narrow-band signal detection, initial broadcasting signal and transmission signal shaping. Since COGSIP is a PAN system, it uses energy detection owing to its low processing load, and the authors introduce method of sensitivity increased by overlapped FFT. Next, a detection method of narrow band signals using autocorrelation is introduced. It achieves detection of a narrow-band signal whose bandwidth is much smaller than an FFT bin width and which cannot be detected by energy detection. Then, an initial broadcasting signal scheme is introduced, which aims to realize announcement of control information to a newly joining terminal that does not know the frequency utilized. The broadcasting is realized without any fixed-frequency control channel. Finally, a transmission signal shaping technique to realize low PAPR (peak-to-average power ratio) in order to reduce interference on primary signals between the COGSIP utilizing bands is introduced. The technique is a modification of DFT-spread OFDM.

In Chapter 5, a novel dynamic spectrum sharing scheme combining spectrum adaptation and MIMO-OFDM for cognitive radio system is proposed. Given spectrum sensing and channel estimation information by the receiver, an improved model due to transmit power thresholds for SA-MIMO-OFDM system is adopted to achieve spectrum adaptation for unlicensed users. In order to improve spectrum efficiency and reduce computational complexity, a simple power allocation algorithm is also proposed. Simulations indicate that the scheme solves the partial interference problem in Interference Temperature Model. The BER performance of the 4Tx4Rx SA-MIMO-OFDM system is over 7 dB higher than the SISO system without spectrum sharing. Moreover, power threshold reduction has little impact on the BER performance of the proposed system with different antenna numbers. Compared to SISO system using Power Threshold Model, the spectral efficiency of the proposed system with 2Tx2Rx MIMO increases 7.8 bps/Hz at a target SNR=30 dB, and the loss in spectral efficiency with different power threshold reduction is at most 1.5 bps/Hz at average SNR of 30 dB.

The purpose of Chapter 6 is to report very recent results on policy-based self-management towards self-adaptive and Cognitive Radio Systems (CRS). The focus will be not on self-* properties of cognitive radio systems but on self-learning policy and cognition; the chapter shows that self-* properties appear as the result of cognition. The chapter will elaborate on radio domain-specific cognition model and on the role of context and policy in that model; specific emphasis will be given to policy multiplexing (this is where a policy disappears becoming embedded in to the fabric of other functionalities), process correctness (cross disciplinary approach based on information modelling and theory, control theory) as a part of assessment of CRS (like a "Turing test" for cognitive radio), and emerging design methodology of CRS.

Large parts of the assigned spectrum is underutilized while the increasing number of wireless multimedia applications leads to spectrum scarcity. Cognitive Radio is an option to utilize non-used parts of the spectrum that actually are assigned to primary services. The benefits of Cognitive Radio are clear when used in emergency situations. Current emergency services rely much on the public networks. This is not reliable in emergency situations, where the public networks can get overloaded. The major limitation of emergency networks is spectrum scarcity, since multimedia data in the emergency network needs a lot of radio resources. The idea of applying Cognitive Radio to the emergency network is to alleviate this spectrum shortage problem by dynamically accessing free spectrum resources. Cognitive Radio is able to work in different frequency bands and various wireless channels and supports multimedia services such as voice, data and video. In Chapter 7, a reconfigurable radio architecture is proposed to enable the evolution from the traditional software defined radio to Cognitive Radio.

Cognitive radios are intelligent wireless systems that learn about their environment and adaptively optimize their performance by modifying their transmission parameters. The promising prospects of cognitive radio to be an advancement in the field of wireless networking and a solution for the spectrum under-utilization problem has made it an independent research discipline on its own. The dynamic and adaptive nature of the radio has many challenges in terms of practical realization. In Chapter 8, an overview of the cognitive radio technology and the latest research developments in different layers of the protocol stack are given. The important topics such as wideband spectrum sensing, transceiver architecture and dynamic spectrum allocation are discussed in detail.

As discussed in Chapter 9, cognitive radio is regarded as a radio that can sense, aware, learn and adapt to the outside environment. Evolutionary algorithms have been shown as a promising technique for adaptation in this book. However, their applications in cognitive radios do not restrict to adaptation at all. The authors have now been working on applying evolutionary algorithms to solve other problems in cognitive radios.

In: Cognitive Radio
Editors: P.D. McGuire et al, pp. 1-36
ISBN: 978-1-60876-604-8

Chapter 1

A SURVEY OF INTERFERENCE MITIGATION IN UNDERLAY COGNITIVE RADIO NETWORKS

Ayman Elezabi[1,a]***, Mohammed Kashef***[2,b]
and Mohamed Abdallah[2,c]
[1]American University in Cairo, Egypt.
[2]Cairo University, Egypt.

1. Abstract

In this chapter we survey the three primary approaches in cognitive radio; namely the interweave, overlay, and underlay approaches. We then focus on underlay secondary networks which have several advantages over the other two. For example, underlay networks avoid the error-prone sensing problem and the need for dynamic re-allocation of spectrum associated with the interweave approach. They also do not require the large amount of side information about the primary users needed in the overlay approach. However, conventional underlay networks have to adopt a conservative approach in transmit power levels and, therefore, interference suppression is needed. After an overview of interference mitigation in underlay networks for ultra-wide band (UWB) systems we present a novel interference-minimizing code assignment (IMCA) scheme for code-division multiple-access (CDMA) secondary networks. While technically an underlay approach the IMCA scheme opportunistically assigns the secondary user codes available in the code subspace to minimize the mean square crosscorrelations

[a] E-mail address: aelezabi@aucegypt.edu
[b] E-mail address: m_t_kashef@yahoo.com
[c] E-mail address: mabdalah@gmail.com

between codes of existing and arriving users. Furthermore, the scheme does not require knowledge of the codes or symbol epoch of the primary CDMA network. It thus attains the advantages of opportunistic and underlay networks and avoids the large side information requirements of overlay systems. Significant performance gains are achieved over random code assignment. We then briefly discuss relaying as another technique for extending the range of underlay networks. Finally, we present our conclusions and assessment of underlay networks.

2. Introduction

Cognitive radio is an area of tremendous opportunity for improving the spectral efficiency of wireless communication systems by allowing secondary wireless devices to exploit under-utilized primary spectral allocations without disrupting primary user communications. There are three main types of cognitive radio networks; namely underlay, overlay and interweave, more commonly called opportunistic, networks. The underlay approach allows concurrent primary and secondary transmissions by requiring secondary signals to limit their interference to the primary users below their noise floor. Overlay schemes require knowledge of the primary networks' codebooks and, possibly, messages to support the primary communications and piggyback secondary transmissions. Opportunistic cognitive radio networks sense unused time-frequency slots to use them for secondary transmissions. In this chapter we survey the three classes and then focus on underlay secondary networks and the interference mitigation techniques applicable to that class. Underlay networks have some advantages over the other two types. For example, underlay networks avoid the sensing problem altogether and the need for dynamic re-allocation of spectrum associated with the opportunistic approach. They also do not require the large amount of side information about the primary users needed in the overlay approach. However, conventional underlay networks usually adopt a conservative approach in transmit power levels. Hence, interference mitigation techniques are very beneficial to the performance of underlay networks. The interference mitigation techniques for underlay networks in the literature are typically proposed for Ultra-Wide Band (UWB) systems and are applied at the receiver. After an overview of these techniques we present a novel interference-minimizing code assignment (IMCA) scheme for code-division multiple-access (CDMA) secondary networks that offers significant improvements in underlay network spectral efficiency. The scheme is cognitive in the sense that it selects transmission parameters based on the received signals. While technically an underlay approach the IMCA scheme opportunistically assigns the secondary user codes so as to minimize the projection of the active primary users' code subspace on the candidate secondary code. This may be likened to 'sensing' the unused, or underused, code subspace 'slots' and filling them except that the 'sensing' in this case is soft. Furthermore, the scheme does not require knowledge of the codes of the primary CDMA network. It thus attains the advantages of opportunistic and underlay networks and avoids the large side information requirements of overlay networks. At the receiver, we present a Modified Minimum Output Energy (MMOE) detector. Finally, we briefly overview the use of relays in underlay networks and present our conclusions and some insights.

3. Underlay, Overlay, and Opportunistic Cognitive Radio Systems

The term Cognitive Radio was first introduced by Mitola [1],[2]. He suggested full cognition of the radio within its environment, of which spectrum awareness is a subset. Spectrum sensing cognitive radio would result in increased spectrum efficiency by, for example, enabling spectrum pooling. The idea of spectrum pooling is that operators owning licensed spectrum that is in some way underutilized would create a pool of their spectra to jointly bring their idle resources to market. The idea has since evolved in many ways and has spawned a large body of research as partly outlined in this chapter. Cognitive radio, or next generation networks, may be divided into three categories: underlay, overlay, and interweave (opportunistic) networks. The underlay approach allows concurrent primary and secondary transmissions [3], [4] by requiring the secondary signals to operate at transmit powers that result in interference levels below the acceptable noise floor for the primary users of the spectrum [5]. This forces underlay systems to operate very conservatively, especially when no information is available about the primary transmissions. The opportunistic approach in cognitive radio is where spectrum sensing occurs followed by spectral allocation for transmission by secondary users in sub-bands that are perceived to be empty. Obviously, the detection process suffers from both a probability of not detecting an active primary user or mistakenly marking a sub-band as occupied by a primary user, in addition to the hidden node problem, particularly in the case of ad hoc structures [6]. In fact, it has been reported that current sensing algorithms, mostly based on energy detection, are unsatisfactory for practical systems [7]. Furthermore, the spectrum sensing must continue periodically to detect the event of primary users arriving in the network. This requires dynamic re-allocation of spectrum and secondary users will regularly change their transmit frequencies. This may be burdensome in fast changing spectral conditions and/or highly mobile primary users. Oftentimes, only highly fragmented bands are available imposing constraints on modulation and other transmission parameters Hence, the underlay approach has some advantages over the opportunistic approach. The overlay approach also allows concurrent primary and secondary transmissions. The common mode of operation for overlay systems is that the secondary users use part of their resources for their own secondary communications and the remainder to relay primary transmissions. This requires a lot of information about primary transmissions, which usually means the primary network must cooperate with the secondary network. The underlay is advantageous over the overlay approach in that it does not need this information or cooperation from the primary network. In the following subsections we further detail each type of cognitive radio approach.

3.1. Overlay Cognitive Radio Systems

The overlay cognitive network exploits the knowledge of the primary users' code-words and the negotiating messages between the primary users to be able to concurrently transmit signals with the primary users. The overlay system can use this information mainly in two ways. First, this information can be used to completely cancel the interference due to the primary signals at the cognitive receiver by sophisticated techniques like dirty paper coding [8]. Second, the cognitive users can utilize this knowledge and assign part of their power for their own communication and the remainder of the power to assist the primary transmissions. By accurate calculation, the cognitive system is able to split its power to enhance the signal-to-noise ratio (SNR) of the primary user by the fraction that is needed to bring the performance of the primary network to its level in the absence of a secondary network.

The overlay concept assumes the ability to know the primary message in a very short time or knowing it prior to its transmission which is not always feasible. The first option of decoding and re-encoding the message during the transmission time of the message needs very high processing abilities. The second option, which is the knowledge of the message prior to the primary transmission, occurs in the case of a primary message retransmission where the cognitive system was able to decode the message while the primary receiver was unable to do so. This second option may also occur if the primary user has the ability to transmit its message to the cognitive user prior to the regular primary transmission in a separate channel. That happens if this separate channel is available and permits this transmission with low power and bandwidth assuming there is a protocol that organizes this operation. Different encoding schemes could be used for the overlay approach. The most common encoding scheme is rate-splitting and it was widely used previously in interference channels, which are channels where different users transmit without any knowledge about the others messages [9]. Rate-splitting is a technique where the cognitive receiver is able to decode part of the primary transmitter's message to be able to cancel the interference caused by that part. As a result the cognitive receiver wastes part of its resources in the decoding process which leads to less transmission rate. This technique was first applied to overlay networks in [10][11][12]. Other techniques could be used to exploit the knowledge of the primary user messages. One of those techniques is referred to as cooperation, or cooperative communication, where the cognitive user is able to decode the primary user message and use a fraction of its power to increase the rate of the primary user. Another encoding scheme is precoding. Precoding is the ability to use the information known about the interference from the primary network to increase the allowed rate in the cognitive network. This information may be the spreading code, the sent messages and the used time slots. A precoding technique referred to as the Gel'fand-Pinsker (GP) binning [13] as well as dirty-paper coding [8] in Gaussian channels may be applied.

3.1.1. Cooperation between Nodes

Cooperative techniques allow nodes to relay each other to increase the network capacity. This type of cooperation requires the relay node to have information about the source message. This information could be obtained from the received signal, by listening to the channel, or by knowing part of the transmitted signal [14]. The simplest communication scenario with cooperation is the relay channel [15]. In cooperation with the source, the relay nodes can decode and forward. In this scheme, the source transmits at the rate such that the relay can decode the message, re-encode it, and transmit it together with the source. Alternatively, the relay can employ compress and forward or amplify-and-forward to respectively quantize or amplify the received signal and send it to the destination.

There are various forms of cooperation between the cognitive and primary users which are:

Oblivious primary users: In this type of cooperation the primary users are oblivious to the presence of the cognitive users. The cognitive users decode the message from the primary source without any assistance from the source.

Aware primary users: In this form the primary users are aware of the presence of the cognitive users. As a result, they will use a decoding scheme in the primary receivers that exploits both transmissions from the cognitive and primary transmitters.

Cooperative primary users: In this setting primary users may assist in delivering data to cognitive users. That may happen if a primary encoder may know channel conditions at a cognitive user. Based on this, it can transmit at a rate which ensures that the cognitive user can decode its transmitted message.

Full cooperation and cognition: This is the most general form where all the cognitive and the primary users are aware of the environment. They all cooperate using the available information to maximize the benefit from both networks.

The messages used for cooperating between the cognitive and primary users should use as little resources as possible. As a result, the cooperation happens usually between the nodes near to each other.

A more detailed discussion of connections between cognition and cooperation can be found in [14]. From the above discussion the complexity of overlay cognitive radio systems, and their interrelationships with primary networks, is apparent.

3.2. Interweave Cognitive Radio Systems

The interweave concept was perhaps the original and most intuitive proposal in cognitive radio [1]. It exploits the idea of opportunistic use of spectrum. The spectrum space-time-frequency holes are voids where there is no activity in a portion of the spectrum at a given time at a certain location. Interweave cognitive users require activity information of the primary users' spectrum to be able to apply the interweave concept in their transmission. This concept is different from the other two schemes as it aims to totally avoid the multiple-access interference to or from the primary network [16].

The basic feature in the opportunistic cognitive system is the detection of the primary signals. This sensing should be accurate and is done for a wide bandwidth which makes it a very complex task. Other source of difficulty in sensing is the primary signal degradation due to fading and shadowing effects. In addition, due to the dynamic activity of the primary users the sensing should be done periodically to update the activity information at the cognitive network [17][18], [19], [20], [21].

One of the detection problems is the low SNR of the received signal at the cognitive receiver. This low SNR requires longer sensing times especially in energy detectors. Even with infinite sensing time, the detection may be impossible when the signal SNR is below a certain limit [17]. Cyclostationary detectors are another more advanced type of detectors but it also has SNR limitations in the detection task [18],[22]. A brief overview of the different sensing techniques is mentioned in the following.

Matched filter detection: This is the optimal detector when the primary information is available to the cognitive user and in the white noise channels [23]. It is optimal as it maximizes the SNR at the detector in the white noise. The information needed about the primary user is its modulation type and pulse shape.

Energy detection: This is the simplest sensing method and the most commonly used type in cognitive radio networks. In energy detection schemes, the energy at a certain frequency band is measured and then a decision on the presence or the absence of the primary user is made [24]. Shadowing and multi-path fading factors are considered for the energy detector in [25].

Cyclostationary feature detection: Modulated signals generally have certain periodic features as they are usually coupled with sine wave carriers, pulse trains, spreading sequences, hopping sequences or cyclic prefixes. These modulated signals are characterized as cyclostationary since their mean and autocorrelation are periodic. These features are detected by analyzing a spectral correlation function. The cyclostationarity detectors compute the spectral correlation function as in [26][27], [28].

Cooperative detection: One solution to combat fading and noise uncertainty is, fundamentally, to take a collaborative approach to sensing [17], [19], [20], [21], [22]. Multiple cognitive users have to independently monitor primary user activity and then exchange spectrum availability estimates to infer the presence of the primary users. These individual detectors may be any type of the above three types of detectors.

The main drawbacks of the interweave cognitive approach are the complexity and uncertainty in the sensing, the overhead for transmission of the sensing information, the dynamic reallocation of spectrum, and the requirement to use fragmented spectrum.

3.3. Underlay Cognitive Radio Systems

The underlay system has a constraint on the power of the secondary signal received at the primary receiver. This constraint is converted to a constraint on the transmitted power [29] assuming knowledge of all channel gains or assuming the ability to estimate the channel using the received signal from the primary users.

Underlay system can exploit multiple transmit antenna techniques to improve the capacity and realize diversity gains in the secondary network in addition to other MIMO techniques including space time coding. In particular, beamforming techniques can be employed where the transmitted signal can be steered to maximize the signal at the secondary network receiver while minimizing the interference to the primary user receiver(s). As a result, the received signal at the primary receiver is within the acceptable interference power limit [30], [31].

As mentioned above, the underlay system basically needs a very large bandwidth to be able to operate at an acceptable SNR with a severe constraint on the power spectral density level. As a result, some form of spectrum spreading technique is needed. The two commonly proposed techniques for underlay systems are Ultra Wide Band (UWB) and direct-sequence CDMA with UWB being the more commonly proposed of the two. The underlay system for cognitive radio networks was proposed using UWB signaling in, for example, [32], [33], [34]. On the other hand, CDMA underlay systems were proposed in [35], [36], [37].

3.3.1. Advantages of Underlay Networks

Under the limited power constraint imposed on the underlay system, it is most probably useable for short range communication systems. Despite this reality, it may be sometimes preferred to use underlay cognitive approach instead of the more common opportunistic approach.

There are many situations where the underlay approach seems advantageous. For example, the underlay system is preferable if some primary users have a frequency hopping signal and/or the spectral conditions are changing very fast. As a result, the cognitive radio may not able to keep track of the spectrum opportunities. The same applies when the primary users are time hopping. Large systems where user arrivals and departures are very frequent result in frequent spectral re-allocations on opportunistic systems which clearly increase the system overhead. One of the most common cases is the situation where the primary user is highly mobile; it may be risky to use the spectrum opportunistically because the communication of the primary system can be easily disturbed. The bands open to opportunistic usage may be too fragmented by narrowband systems to the extent that it would be undesirable to apply the opportunistic approach due to the difficulty of both the sensing problem and utilization of spectral holes.

The underlay approach is also preferable when there is a high level of uncertainty in the detection information for the spectrum hole. As an example, if the spectrum sensing results from different nodes in a cognitive network do not match to a large extent, this may indicate that either some of the nodes are being shadowed, or the spectrum sensing information of some nodes is unreliable. It should also be mentioned that transmitting the sensing results and guaranteeing their integrity constitutes a bandwidth and complexity overhead on secondary networks.

In conclusion, underlay networks avoid the sensing problem altogether and its associated drawbacks. The underlay approach is also preferable over the overlay approach as it does not require any explicit side information or cooperation from the primary system.

3.3.2. Challenges in Underlay Networks

The underlay system capacity maximization problem is formulated as a general multiuser communication problem with transmit power constraints at the cognitive transmitters that determines the performance of the cognitive network.

Underlay system, by definition, must use very limited power levels in their transmissions as mentioned earlier. The interference power from the primary network is therefore very high relative to the received power at secondary nodes in underlay systems. This limitation on the transmitted power creates constraints on both data rate and communication range.

To enhance the performance of underlay systems, there are two main directions. The first is to use a very wide bandwidth and the second is the application of interference mitigation techniques. Typically, a combination of both is needed to make underlay systems feasible. Using a large bandwidth makes the secondary network coexist with a larger number of primary users as the cognitive signal in this case spans a large number of narrowband primary users. On the other hand, applying interference mitigation techniques may require knowledge or side information about the primary transmissions that is obtained by advanced signal processing at the transmitter and/or receiver. The main challenges in using cognitive underlay networks are the ability to coexist with different primary networks, even with very large secondary bandwidth, and the ability to limit the interference from the primary network. The following section is dedicated to interference mitigation in underlay networks.

4. Interference Mitigation in Underlay Networks

Underlay systems are typically either Direct Sequence CDMA (DS-CDMA) systems with high processing gain or UWB systems. As mentioned earlier, the transmitted signals are characterized by having low level PSD which leads to the problem of being affected by the primary narrowband interferers with higher PSD level. To alleviate this problem and improve the performance of the underlay systems interference mitigation techniques are needed. Such techniques would make underlay systems feasible in practical systems and extend their range and/or enable higher data throughputs. Interference mitigation techniques are also used in non-underlay systems. The main difference is that comparable power levels are expected between the desired signal at the secondary receiver and the interference signal. Also, both signals usually have similar bandwidths. Certain interference mitigation techniques are, therefore, particular to underlay networks. In this survey we loosely classify underlay systems into cognitive and non-cognitive ones. First, we discuss the non-cognitive underlay systems where there is no information about the primary system that is used by the secondary network transmitters and hence there is no attempt to change the transmission parameters of the secondary system. In that case, any interference mitigation is performed at the receiver. Next, we explore interference mitigation techniques in cognitive underlay systems where the secondary network attempts to select the users' transmission parameters based on some cognition of the received signals in the surrounding environment.

4.1. Interference Mitigation in Non-cognitive Underlay Systems

4.1.1. Filtering

In this section, we overview the filtering approach for interference mitigation in non-cognitive underlay systems which is considered the most widely used method. We then present a brief survey of interference mitigation techniques in the UWB systems based on the filtering approach.

The main concept behind the various filtering schemes is the ability to detect the band that the narrowband interference occupies in the spectrum of the secondary signal so that the secondary system is able to filter that interference out. In the following subsections we overview different methods for designing these filters.

4.1.1.1. Linear Prediction Filters

The filter's main function is to whiten the received signal by making its samples uncorrelated to achieve the goal of removing the narrowband interference. It depends mainly on the fact that the wideband signal samples are uncorrelated while the narrowband interference signal samples are highly correlated [38], [39]. Hence, the filter is able to distinguish these correlated samples and remove them.

The linear prediction filters are commonly implemented by a tapped delay line where X_j is the samples of the received signal, a_i is the filter coefficients and y_j is the output of the filter for the jth sample.

The filter generates the interference and then subtracts it from the received signal where the estimated interference can be calculated by the equation

$$\hat{I}_j = \sum_{k=-N,k\neq 0}^{k=N} a_k X_{j-k} \tag{1}$$

The coefficients of the filter are selected to minimize the mean square error $E[y_i^2]$ where y_i is the output of the filter after interference cancellation. This is the well known Weiner-Hopf filter. To have fixed coefficient values, the correlation function of the received signal should be known to the receiver [40]. In practice, that is not feasible as the correlation function is time varying. As a result, the filter coefficients are calculated adaptively.

The most widely used adaptive algorithms for calculating the filter coefficients is the Least Mean Square (LMS) algorithm [38]. The adaptation equation is

$$w_{j+1} = w_j + \mu x_j \varepsilon_j^* \tag{2}$$

Where w_j is the vector contains all the filter coefficients such that $w_j=[a_{-N}\ a_{-N+1} \ldots . . a_N]^T$, μ is the adaptation parameter and ε_j is the error in the jth sample between the filter output and the decision made. Different adaptive algorithms were considered and compared in the literature [41], [42], [43].

4.1.1.2. Non Linear Prediction Filters

The linear filter is designed based on the assumption of that the DS-CDMA signal is Gaussian. However, it was found that the DS-CDMA signal is a non-Gaussian process to the filter [38][39]. As a result, the linear filters are suboptimal and the non linear prediction filters can be employed to achieve better performance. It is found that the optimum non linear filter is a series of Kalman-Bucy filters [41]. However, the implementation of such filters suffers from high complexity [44]. Therefore, an Approximate Conditional Mean (ACM) filter was proposed [44] where its coefficients are obtained using adaptive algorithms similar to the linear filter case. In particular, the LMS algorithm is also used in the non linear filters where the current value of the coefficients is given by a linear function of the previous value, the same as linear filters, and is modified by a non linear function of the error value.

Another gradient algorithm that may be used in the adaptation process is the non linear gradient algorithm. It gives better performance than the linear gradient LMS algorithm in some situations but is much more complex. It also has the problem of having a number of local minima that it can undesirably converge to resulting in poor performance. The linear gradient on the other hand has only one global minimum [44].

One other non linear filter is the decision feedback filter [45]. It depends on using the previous decisions in the filtering process. It suffers, however, from the problem of error propagation in the case of low Signal to Interference power ratios.

4.1.1.3. Transform Domain Processing

In this type of processing, converting the received signal into the frequency domain is first carried out then the required filter is applied to the received signal [46]. A notch filter is then used to remove the narrowband interference. Another way is to limit the power of the received signal in the affected part of the spectrum by attenuating partially this part such that a small part of the narrowband interference power remains in the signal. This way is different from the notch filter such that it keeps most of the underlay system power with a certain amount of the narrowband interference power while the notch filter attenuates both the signal and the interference in a certain portion of the spectrum.

4.1.2. Other Interference Mitigation Techniques

There are some interference mitigation techniques primarily related to UWB synthetic aperture radar. In these techniques, the RELAX algorithm, which is described in [47], is used in sinusoidal estimation. In this algorithm, the interference is assumed to be a group of weighted complex exponentials. The exponentials frequencies and weights are estimated to minimize the error in the interference estimate. Another interference suppression technique proposed in [48] uses LMS algorithm in interference estimation. The interference is estimated by multiplying the received signal by a certain matrix derived from the equation of the mean square error between the interference and the received signal where the error in this case represents the required message.

4.2. Interference Mitigation in Cognitive Underlay Systems

We now discuss interference mitigation in the cognitive underlay UWB systems where the transmitted signal is adjusted cognitively. The Detect And Avoid (DAA) technique is mostly used in pulse-based UWB and Multi-Band UWB (MB-UWB) [49]. In this technique, the narrow band primary users are detected and then using a certain spectrum shaping technique, the secondary system avoids this part to enhance the secondary system performance. In [50], an interference mitigation technique for Direct Sequence Multi Band Ultra Wide Band (DS-MB-UWB) system was proposed. In the DS-MB-UWB, each user has a spreading code to select the frequency sequence of the sub-bands to be used. By partially suppressing the affected sub-band, the effect of the narrowband interference is mitigated while maintaining the full frequency diversity. Narrowband interference detection is done based on energy detection where a bank of band-specific energy detectors also determines the sub-band affected by the interferer. Depending on the energy of the narrowband interference, the suppression coefficient of the frequency sequence is determined to minimize the energy in the affected sub-band where the system reduces the probability of using the affected frequency band. Also in [51], sculpting for MB-OFDM spectrum to avoid overlapping with the WiMax users is used. Sculpting here refers to reshaping the UWB signal spectrum to minimize the effect on the WiMax system. The same technique is used in [52] where The MB-OFDM transmission scheme is used and the detection and avoidance technique is used to mitigate interference at the sub-bands affected by narrowband interference.

In [53], the interference mitigation function within a network is performed by a central node by deciding the waveform shape that could be used by each user.

Adaptive pulse shape waveforms are investigated and used in pulse-based UWB to suppress a certain band where the narrowband interference exists [54], [55]. A multicarrier type template wave was used to mitigate the interference of IEEE802.11a signal on the UWB signal [56]. The waveform of the UWB pulse is selected to suppress the transmitted signal at a certain frequency band. It is also proved to be effective in the MB-OFDM interference to the UWB system. This is achieved using a multicarrier template that uses different waveforms in the different carriers [57]. This technique can be also applied to the DAA technique as in [49]. Applying a spreading sequence for UWB in the frequency domain is also used in interference mitigation, which places a null at the band of the narrowband interference signal. This technique needs to know the narrowband interference frequency and uses Fourier transform in encoding the signal in the frequency domain [58].

For the less common type of DS-CDMA underlay networks, to the authors' knowledge, no works on interference mitigation appear in the literature. In the following section we present a novel cognitive underlay interference mitigation scheme for DS-CDMA networks. It is based on a code assignment strategy that minimizes interference to the secondary network. At the secondary receiver, a modified Minimum Output Energy (MOE) detector is proposed.

5. [1]CDMA Cognitive Underlay System

In this section we describe an underlay system using Direct-Sequence Code-Division Multiple-Access (DS-CDMA) [35], [59]. We consider CDMA signaling in both primary and secondary networks and, for convenience, assume equal spreading factors in our derivations. However, the methods introduced are extensible to primary and secondary CDMA networks with unequal spreading factors [60] and, in principle, to other primary networks as long as some periodicity is exhibited in their transmissions. The basic premise of underlay systems is that the interference to the primary network is negligible by limiting the secondary network transmit powers. This may be compensated for by having lower data transmission rates and higher spreading factors for the secondary users.

---

[1] This Section is adapted from 'A. Elezabi, M. Kashef, M. Abdallah and M. Khairy, "CDMA Underlay Network with Cognitive Interference-Minimizing Code Assignment and Semi-Blind Interference Suppression," Journal on Wireless Communication and Mobile Computing, *vol. 9, issue 11, November 2009*, DOI:10.1002/wcm.724' with modifications.

Hence, the interference problem we deal with is that from the primary to the secondary network as well as the Multiple-Access Interference (MAI) within the secondary network itself.

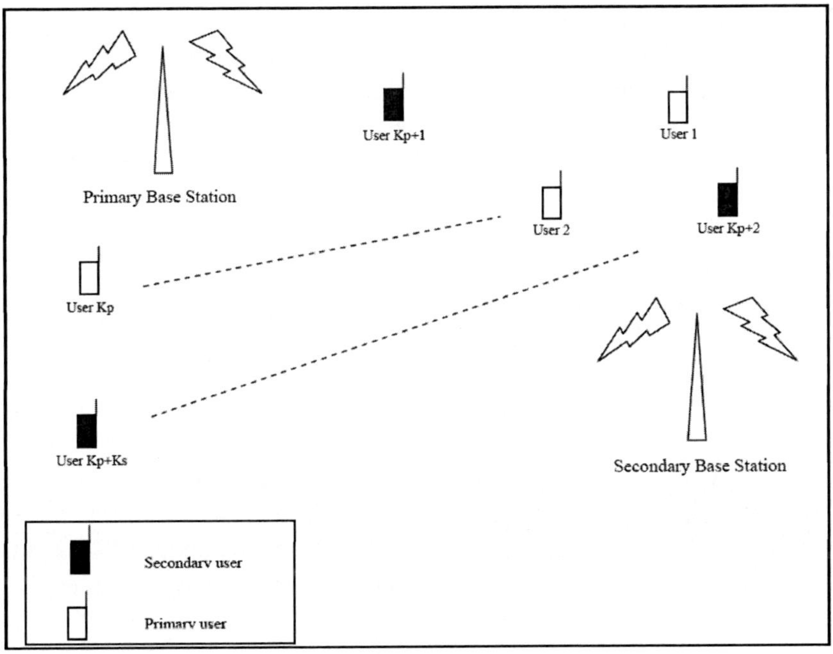

Figure 1. Proposed Cognitive network scenario.

We assume an infrastructure based framework where a secondary network base station applies interference mitigation techniques to improve secondary network capacity. No information is passed from the primary network to the secondary network base station. This network structure is shown in Figure 1.

In order to minimize the interference from the primary CDMA network we propose a novel cognitive method that we call interference-minimizing code assignment (IMCA) for underlay secondary networks. Without cognition, underlay CDMA systems would require unreasonably high chip rates. On the other hand, IMCA may enable DS-CDMA as a possible technology for underlay networks. The idea is based on assigning spreading codes to each newly arriving secondary user such that the interference due to existing primary and secondary users experienced by that user is minimized. This is accomplished by measuring the covariance matrix for the received signal from the existing primary and secondary user transmissions and selecting the code that minimizes the mean square cross-correlation between the received signal and the candidate codes. Since the primary user spreading codes are unknown to the secondary network we adopt a blind approach and formulate the problem as a constrained minimization problem over the space of available codes that is equivalent to minimization of a Boolean quadratic form. The solution may be obtained exhaustively but has exponential complexity in the spreading factor N. We therefore reduce the multiple constraints on the user code space to a single constraint and obtain a near-optimal Lagrangian solution with linear complexity.

The performance using this near-optimal method is found to be very close to that when using codes obtained from the exact solution. The performance gains due to IMCA measured in terms of error probability compared to Random Code Assignment (RCA) are impressive. Note that while the original goal was to minimize the effect of the primary user transmissions on the secondary users, based on the premise that the underlay network is operating beneath the acceptable noise floor of the primary network, the IMCA scheme will inevitably result in further reducing the interference caused by the secondary network onto the primary network. This may naturally be translated to higher throughput for the secondary network. When comparing between typical cognitive, i.e. spectrum sensing, radio networks and the proposed IMCA underlay networks an analogy may be drawn to the comparison between FDMA/TDMA and CDMA systems in scheduled communications. However, for secondary network operation the advantages are expected to be far greater. For example, as mentioned earlier the IMCA scheme does not suffer from sensing errors that result in either misallocated or under-allocated spectrum. Furthermore, no dynamic re-assignment of spectrum is needed. The base station will re-assign an existing user with a new spreading code on an as-needed basis. The IMCA scheme is described in this work for synchronous primary networks using short spreading codes where the primary user bit epoch is available to the secondary network. We consider asynchronous primary networks in [36] where the Lagrangian optimization problem is modified and a scheme for blind epoch acquisition is proposed. Similar performance gains are achieved with IMCA in chip-asynchronous multipath fading channels.

For the receiver at the secondary network base station receiver a logical choice would be to use the minimum-output energy (MOE) detector, which does not require knowledge of any user codes other than the user of interest. A similar problem is that of inter- and intra-cellular interference suppression in cellular CDMA systems where a base station has knowledge of the signature sequences of the users it is serving but not those of interfering users in co-channel cells. The usage of the knowledge of the secondary users' spreading codes in such cases was proposed as semi-blind approaches for inter-cell interference cancellation [61][62]. However, in cellular environments the proposed schemes operate on the basis that the users of the known spreading codes, i.e. those served by the base station, have higher received power than users whose codes are unknown. Hence, it is appropriate that the first stage of interference suppression be parallel interference cancellation (PIC) for the users with known spreading codes followed by a MOE detector as in [63]. However, in our case due to the fact that the secondary network base station has no power control over the primary user mobile uplink transmissions, the received powers of those users may be higher than the uplink secondary user transmissions. Hence, we propose applying an adaptive implementation of the Minimum Output Energy (MOE) [64] detector first to improve the quality of the secondary user decisions followed by PIC among the secondary users only. Another advantage of the MOE detector is that it needs no training sequence, which is not available from the primary users. The original formulation of the MOE, however, does not benefit from available knowledge of the secondary user codes. In this work we incorporate such knowledge by using a partial decorrelating filter [65], based on the known secondary user codes, as the seed to the adaptive MOE detector. We show that we obtain faster convergence of the adaptive algorithm compared to the conventional case of starting with the matched filter seed. In fact, we can decrease the time required for the conventional MOE to converge by half.

It is reasonable to assume that our secondary underlay network exercises some form of power control over the secondary mobile users. Hence, we apply PIC between the secondary users since PIC is a suitable choice in that case [66].

5.1. System Model and Receiver Structure

We consider K_p primary users and K_s secondary users transmitting bit-synchronously using binary CDMA signaling over a frequency-nonselective Rayleigh fading channel with additive white Gaussian noise. At any bit interval, the un-normalized received signal vector is given by

$$r = \sum_{i=1}^{K_s} b_i c_i s_i + \sum_{j=K_s+1}^{K_s+K_p} b_j c_j s_j + n \tag{3}$$

where $c_j = | c_j | \exp(i\theta_j)$ is the complex Gaussian fading coefficient for user j, b_j is the bit of user j ($b_j = \pm 1$) using BPSK symbols, s_j is the spreading code of user j

with spreading factor equal to N, and n is the thermal Gaussian noise vector with distribution $N(0, N\sigma^2 I)$. All spreading codes are short codes that repeat every bit. With 1 sample per chip, the signal r is an (Nx1) vector. Equation (3) can be alternatively represented in the following matrix format,

$$r = S_s C_s b_s + S_p C_p b_p + n \tag{4}$$

In the above equation, $S_s = [s_1\ s_2\ s_3 \ldots\ldots s_{Ks}]$ represents the matrix containing all the spreading codes of the secondary users, C_s=diag($[c_1\ c_2\ c_3 \ldots\ldots c_{Ks}]$) is the matrix for the fading coefficients of the secondary users, and $b_s=[b_1\ b_2\ b_3 \ldots\ldots b_{Ks}]$ is the vector of the secondary users' bits. The symbols S_p, C_p, b_p are defined similarly for the primary users.

As the spreading codes of the primary users are not known, a suitable candidate for the receiver of the secondary users should come from the class of blind or semi-blind multi-user detectors [64]. As mentioned earlier we use the MOE detector and apply PIC among the secondary users based on the output of the MOE stage. Figure 2 shows a block diagram of the system.

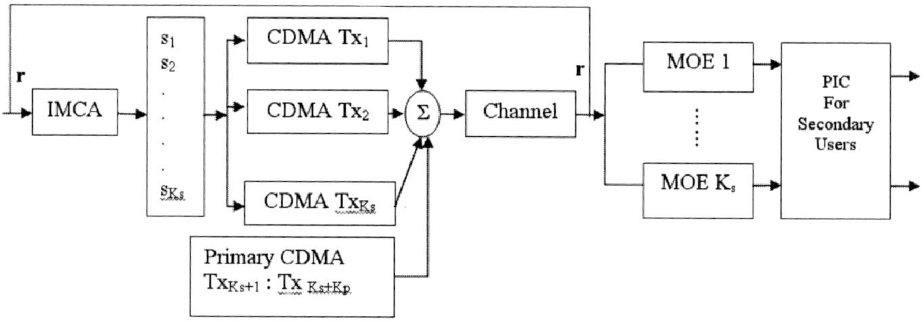

Figure 2. System Block Diagram.

The MOE receiver correlates the received signal r with a vector w that can be calculated for each user as in [64]

$$w[i] = s_1 + c[i] \tag{5}$$

where s_1 is the spreading sequence of user 1, $c[i]$ is the adaptive part orthogonal to s_1, and i is the bit index. The adaptive equation to calculate $c[i]$ as mentioned in [67] is

$$c[i+1] = c[i] - \mu \operatorname{Re}\{z[i]\}\left(r(i) - \operatorname{Re}\{z_{MF}[i]\}\, s_1\right) \qquad (6)$$

where $z[i]$ is the output of the MOE filter, $r[i]$ is the received vector, $z_{MF}[i]$ is the output of the matched filter and $\operatorname{Re}\{.\}$ denotes the real part.

In order to improve the performance of this blind detector using the information of the known codes of the secondary users we propose using the partial decorrelator [65], i.e. using partial knowledge of the user codes, as a seed for the adaptive MOE technique. The conventional MOE includes a vector component $c[i]$ that is orthogonal to the spreading code of the desired user. The adaptation of this orthogonal component is driven to minimize the output energy. The main idea in the improved MOE is that a part of this orthogonal vector is known and can be obtained from the spreading codes of the secondary users. While the adaptive MOE should converge to the filter obtained from the closed form solution $w = (s_1 A^{-1} s_1)^{-1} A^{-1} s_1$, where $A = E[r\, r^H]$ is the covariance matrix of the vector r, the improved MOE is expected to improve the convergence speed of the MOE. In the improved MOE the received data is correlated with the vector w_{dec}, where it is given by

$$w_{dec}[i] = s_{1,dec} + c[i] \qquad (7)$$

where $s_{1,dec}$ is the partial decorrelating vector for user 1 using the known secondary users' codes, and $c[i]$ is calculated in the same way of equation (6). The partial decorrelating filter $s_{1,dec}$ is given by the first row of the matrix $(S_s^T S_s)^{-1} S_s^T$ for user 1, the user of interest. Note that the matrix inversion operation is computed only when secondary users arrive or depart.

5.2. Interference-Minimizing Code Assignment

In this section we describe the IMCA scheme used for spreading code selection in the underlay CDMA network in order to achieve the goal of minimizing interference. Any arriving secondary user is assigned a spreading code such that the mean square cross-correlation between the received signal and that code is minimized. This is accomplished by minimizing an objective function subject to confining the codes elements (chips) to take values of $\pm$ 1. For clarity, we start with the case of a single arriving secondary user with K_p active primary users then discuss the case of multiple secondary users.

5.2.1. First Arriving Secondary User

Let the first arriving secondary user be user 1. Then we seek to assign user 1 an optimal spreading code s_1 where the optimality criterion is to minimize the average square cross-correlation between the spreading code s_1 and the received signal r. In particular, our objective is to find the cognitive secondary user code s_1 that minimizes the following

$$s_1 = \arg\min_{s_1}(E(|s_1^H r|^2)) = \arg\min_{s_1}(s_1^H A s_1) \tag{8}$$

where s_1^H denotes the conjugate transpose of s_1 and A is the covariance matrix of the vector r defined earlier. The minimization is subject to the constraint that the individual component of s_1 takes the value ± 1 or equivalently the individual components of the vector s_1 should satisfy the following relation

$$s_{1i}^2 = 1 \qquad \text{for } i = 1, 2, \ldots., N \tag{9}$$

The process of finding the optimal code s_1 can be obtained using two different methods. The first one is to conduct an exhaustive search on all the possible values of the vector s_1 to find the optimal one. Although, this approach is useful in performing system simulations and assessing the performance, it is not yet practical in real implementation as it suffers from a complexity order that is exponential in N. The second method can be developed by formulating the problem as a Lagrangian optimization problem where the Lagrangian objective function to be minimized is given as follows

$$J = s_1^H A s_1 + \sum_{i=1}^{N} \lambda_i (s_{1i}^2 - 1) \tag{10}$$

where λ_i are the Lagrangian multipliers. The code s_1 may then be obtained by applying the standard method of differentiating the Lagrangian function with respect to the individual components of s_1 and the Lagrangian multipliers λ_i. This results in $2N$ nonlinear equations that must be solved simultaneously to find the code s_1. It is obvious that obtaining the optimal s_1 using this method is computationally tedious as it involves the solution of nonlinear equations. This is equivalent to the well-known problem of minimizing a quadratic Boolean form such as $s_1^H A s_1$. Some methods have been proposed for obtaining near-optimal solutions such as semi-definite programming [68] but they remain quite complex. Hence, there is a need to find an alternative method that can more efficiently find a near-optimal code s_1, i.e. resulting in a near-minimum value for the quadratic form $s_1^H A s_1$.

5.2.2. Near-Optimal Method for Code Assignment

We now develop a near-optimal method that finds the vector s_1 that almost achieves the minimum mean square cross-correlation. This method depends on the observation that since the vector s_1 satisfies the relation (9) then subsequently the vector s_1 must satisfy the following

$$s_1^H s_1 = N \tag{11}$$

Hence, rather than finding the optimal s_1 vector whose elements obey the relations (9) we develop an optimization problem that finds the code s_1 and satisfies the single constraint (11). In particular, we find the code s_1 that minimizes the following Lagrangian function

$$J = s_1^H A s_1 + \lambda(s_1^H s_1 - N) \tag{12}$$

Due to this relaxed constraint we have a single Lagrangian multiplier. It can be easily shown that the eigenvectors of the covariance matrix A minimize the above Lagrangian function where the Euclidean norm of the eigenvectors are scaled to satisfy the relation (11). However, the resulting eigenvectors may not satisfy the relation (9). Hence, we apply a hard limiter, sgn(.) function, to the eigenvectors to make them acceptable as spreading codes and then select from the among them the spreading code that minimizes the objective function J. Note that the number of eigenvectors available is at most N and therefore the search space is among N candidate spreading codes, i.e. complexity $O(N)$. It is clear that this method may not result in finding the same s_1 that results from solving the original optimization problem which is equivalent to what the exhaustive search would produce. However, as our simulation results show, the near-optimal method performance is very close to that of the optimal method in terms of bit error rates for the uncoded system.

5.2.3. Multiple Secondary Users

In the case of multiple existing secondary users in addition to K_p existing primary users. There are two approaches to assigning the j-th arriving secondary user. The first is to solve the same minimization problem of equation (8) without changing the spreading codes assigned to the previous $j - 1$ secondary users. Knowledge of the codes of those previous secondary users is available to the secondary network base station but is not needed explicitly since the received signal and the computed covariance matrix A includes the contributions of those secondary users The second approach is to jointly re-assign all spreading codes for all previous $j - 1$ secondary users whenever a user j arrives. The joint code assignment may be also be formulated as a Lagrangian optimization problem but its solution is very tedious even while using the near-optimal approach as it results in $2N.j$ simultaneous equations to be solved. The exhaustive method becomes prohibitive even for small N since the search space contains $2^{N\ j}$ vectors. Fortunately, we are able to show by simulations that the first approach where no re-assignment of secondary codes takes place for new secondary arrivals is near-optimal.

5.3. Simulation Results

We start by comparing the uncoded error probability performance of the IMCA and Random Code Assignment (RCA) schemes. In all the simulation comparisons we shall assume equal user SNR. As well, the spreading factors for both primary and secondary users are the same for convenience. For the uncoded case the performance is averaged over an ensemble of that varies from 20 to100 randomly generated primary codes with 20,000 bits for each realization. Figure 3 shows the uncoded BER for a single secondary user applying conventional matched filter detection with the proposed IMCA system compared to RCA on a Rayleigh fading channel. The number of primary users $K_p = 7$, and the spreading factor $N = 8$. The IMCA is obtained using the exhaustive and near-optimal methods. As shown in the figure, the performance of the IMCA method is significantly better than RCA method. The near-optimal method gives performance that is very close to the optimum exhaustive search method.

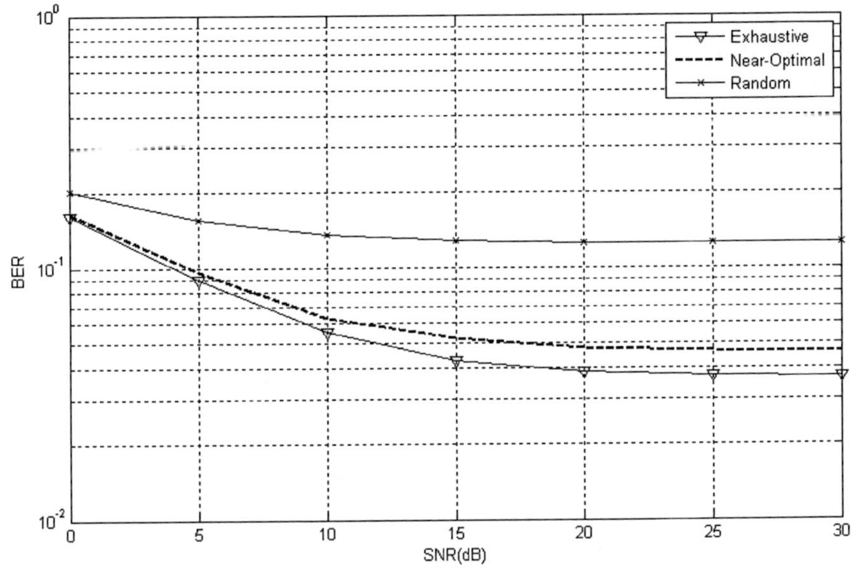

Figure 3. Performance of the IMCA and RCA schemes using MF (Matched Filter) receiver in Rayleigh fading channel. $N = 8$, $K_p = 7$, $K_s = 1$.

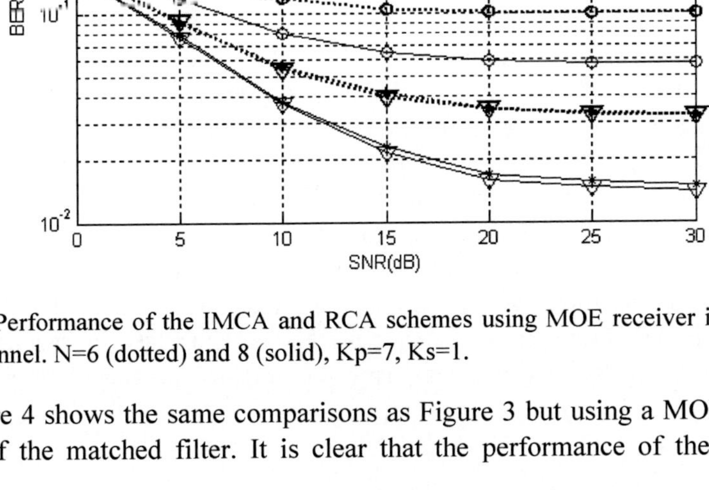

Figure 4. Performance of the IMCA and RCA schemes using MOE receiver in Rayleigh fading channel. N=6 (dotted) and 8 (solid), Kp=7, Ks=1.

Figure 4 shows the same comparisons as Figure 3 but using a MOE receiver instead of the matched filter. It is clear that the performance of the proposed

IMCA scheme is again superior to that of RCA. The near-optimal method is almost identical to that of the optimum exhaustive code searching scheme. Figure 4 also shows the performance for an overloaded system with spreading factor of $N = 6$ and the same number of users. The proposed IMCA scheme still gives superior performance to RCA.

Figure 5 illustrates the value of the normalized objective function J for the same parameters as in Figure 3. It is an indication of the multiple access interference plus noise variance experienced by an arriving user using the assigned code depending on the different assignment schemes. The IMCA system gives good performance over the entire SNR range and the difference between the exhaustive and near-optimal schemes is small.

Figure 6 shows the summation of the squared cross-correlations (SSXC) between the assigned spreading code for user 1 (secondary) and the codes of all other primary users on a Rayleigh fading channel. There is a single secondary user, the spreading factor $N = 8$, the SNR = 10 dB, and the number of primary users is varied from 1 to 11. Hence, the total number of users in the system varies from 2 to 12. The IMCA scheme results in significantly smaller SSXC than in the RCA schemes. Furthermore, the difference between the exhaustive and near-optimal schemes is insignificant even in overloaded networks. Note that the SSXC for the case of RCA may be computed analytically as $(K - 1) / N$ using the standard Gaussian approximation where $K = K_p + K_s$ is the total number of primary and secondary users. The slope of the IMCA curve is less than the slope of the RCA curve which indicates the higher advantage with increasing the number of users. Finally, we investigate the performance of the IMCA when successively assigning several secondary codes. Figure 7 shows the uncoded BER of each arriving secondary users as the number of secondary users is increased. The number of primary users K_p is fixed at 4 and the spreading factor $N = 8$ and the SNR = 10 dB. We note that the performance of the IMCA is again clearly superior to the RCA scheme. Also, the slope of the BER curves is lower for the IMCA indicating the robustness of the IMCA for heavily loaded networks. More importantly, the BERs for all secondary users are similar indicating that secondary users are not disadvantaged due to arriving early or late. It may be perceived that secondary users arriving later face a more constrained code assignment problem. While that is true by itself, secondary users arriving early will suffer from secondary users arriving later whose code assignments they, i.e. the early users, do not determine. Fortunately, the code assignment exercised by secondary users arriving later results in codes that have partially minimized cross-correlations from the perspective of the early users.

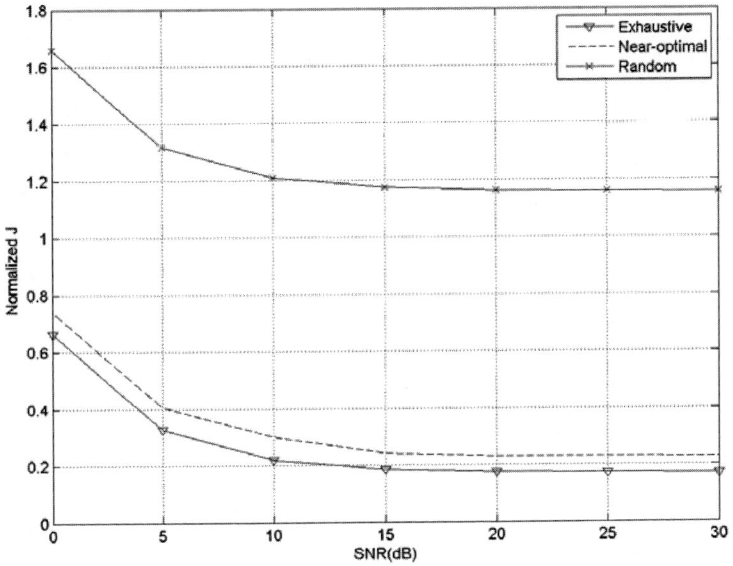

Figure 5. The normalized objective function J in Rayleigh fading channel. $N = 8$, $K_p = 7$, K_s=1.

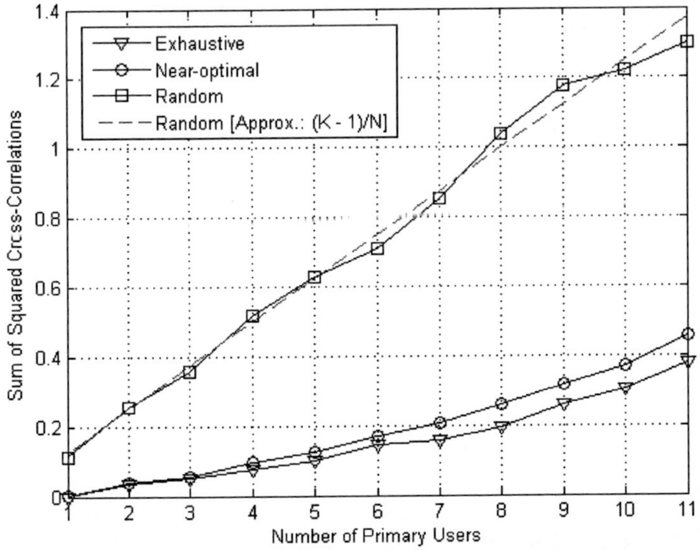

Figure 6. Summation of squared cross-correlations of user 1 (secondary) against number of primary users in fading channels. $N = 8$.

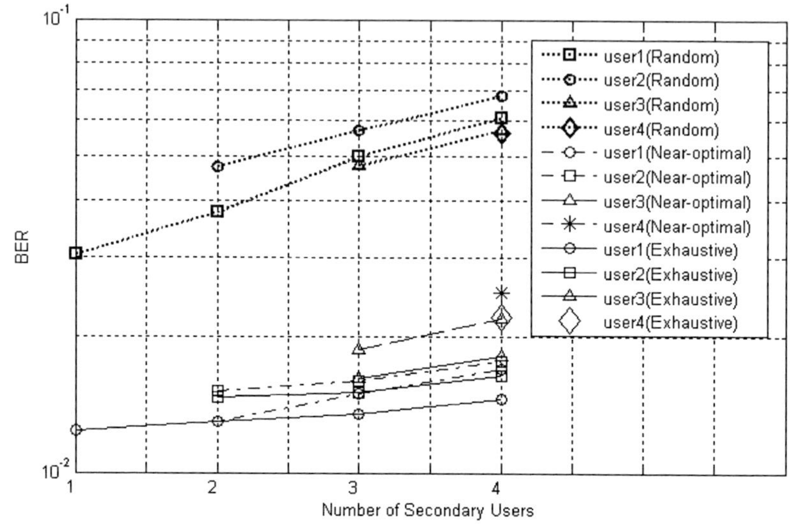

Figure 7. Performance of the IMCA and RCA schemes for successively arriving secondary users using MOE receiver in Rayleigh fading channel. $N = 8$, $K_p = 4$. SNR = 10 dB.

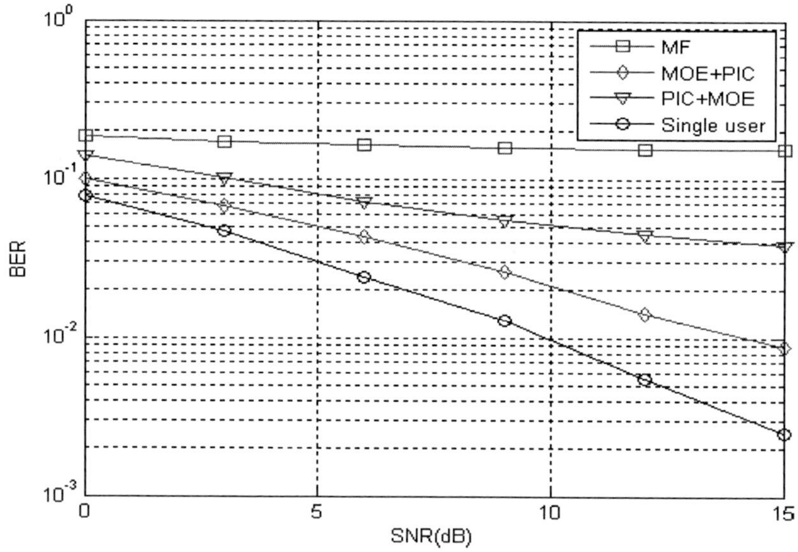

Figure 8. Performance of different receiver structures in Rayleigh fading. $N = 8$, $K_p = 4$, $K_s = 4$. Conv System refers to PIC followed by MOE.

Finally, in Figure 7 the small increase in BER with increasing number of secondary users from $K_s = 1$ to $K_s = 4$ shows that the re-assignment of *all* secondary codes whenever a secondary user arrives will not lead to large improvements. Such a re-assignment would involve huge complexity since all secondary codes would have to be jointly assigned or re-assigned.

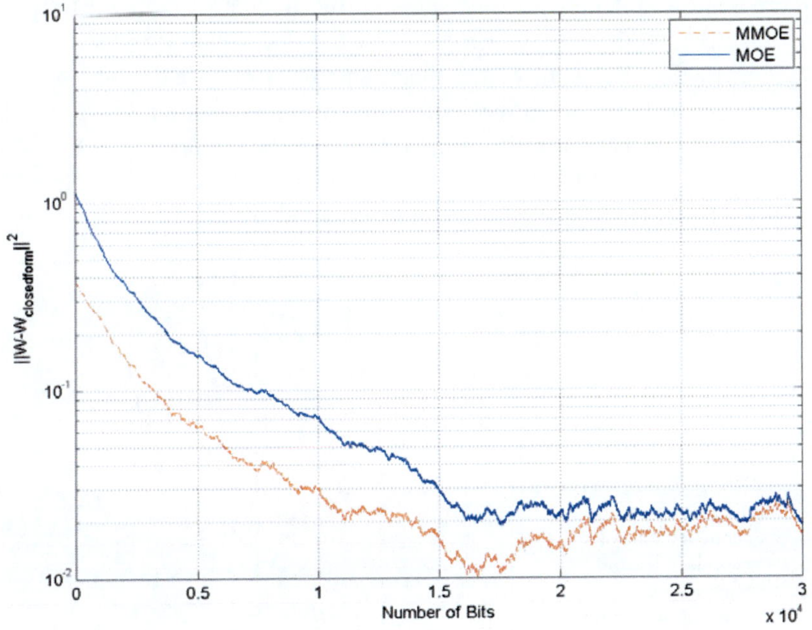

Figure 9. MSE for the MOE and MMOE schemes at SNR=10dB and in the presence of 8 users.

We now demonstrate the justification of applying the MOE prior to secondary user PIC. Figure 8 shows the uncoded performance of the MOE followed by PIC compared to applying secondary PIC followed by MOE in the presence of fading. The matched filter receiver and single-user bound performance are included for reference. The number of primary users is $K_p = 4$ and the number of secondary users is $K_s = 4$ and the spreading factor is $N = 8$. As shown in the figure, the proposed MOE followed by PIC scheme gives a much better performance than the PIC followed by MOE architecture. This is expected since we simulate systems with equal average user energies in this case. The PIC followed by the MOE architecture is used in cellular systems [63] where the PIC is first applied to cancel the intra-cell interference caused by the strong users in the same cell who are closer to the receiver. The MOE receiver is then applied to decrease the effect of the inter-cell interference caused by users in neighboring cells who cause weaker interference. In our scenario, however, the known secondary users and the unknown primary users have the same average power levels and as a result removing the interference first before applying the PIC scheme is better in decreasing the overall interference.

Figure 9 shows the convergence speed of the standard adaptive MOE scheme and the adaptive MOE scheme using the partial decorrelator as the seed in (6). The SNR = 10 dB. The sum of the mean square error (MSE) over the N filter coefficients between the adaptive and closed form implementations, i.e. $\| w - w_{\text{closed form}} \|^2$, is plotted for the standard MOE and modified MOE (MMOE) scheme. The MSE is plotted for each bit transmission. It is clear that the MMOE scheme has a faster convergence speed that would result in a better BER.

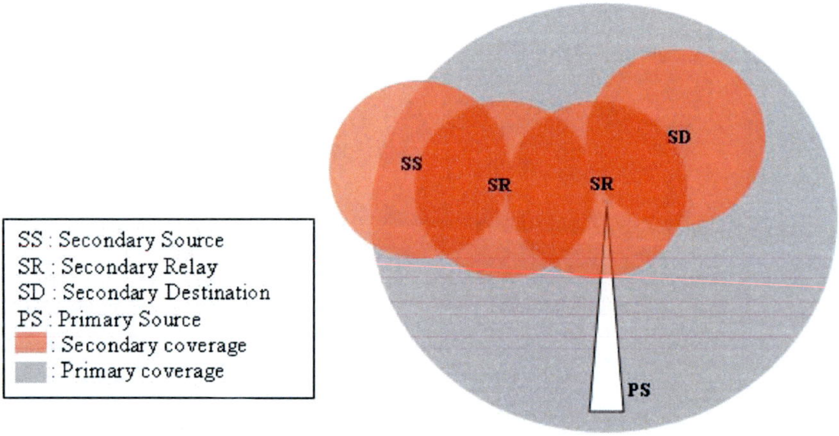

Figure 10. Underlay Relay Network structure.

6. Relaying in Underlay Networks

An alternative approach to improving the performance of the underlay networks is by the use of relaying. Cooperation between relay nodes enables significant SNR and diversity order enhancement and naturally extends the range of secondary underlay networks. The most common scenario is when there is a single transmitter, a single receiver, and a number of relay nodes as discussed in [69], [70]. Figure 10 shows the structure of a relay network where there are a number of secondary users with small coverage areas. Those secondary users in the underlay network use a relaying scheme to extend their coverage areas under the power level constraint. Relays can use in their forwarding the data both regenerative and non-regenerative schemes. The regenerative relays are preferred in short distance cases where the relay is able to decode the data correctly. The relays used may also be fixed or mobile relays. Fixed relays are used when they are placed such that good propagation conditions between them are guaranteed. On the other hand, mobile relays are used when there is a large number of relays available for transmission [69].

7. Conclusion

In this work, we presented an overview of the different types of cognitive radio networks which are the underlay, overlay and interweave techniques. The operational mode of each type and the basic contributions in the literature are briefly mentioned. We then focused on the underlay systems and highlighted its advantages over the other two cognitive radio types. We argued that these advantages make underlay networks a strong candidate for certain applications, e.g. short range communications. The challenges facing the design of the underlay networks were also described.The major challenge in underlay networks is the use of low transmit power levels relative to the primary system. As a result, interference mitigation is one of the most important issues in the design of underlay networks. In this work, we surveyed the interference mitigation techniques that are used in different types of underlay systems. We distinguished between underlay techniques that attempt to mitigate primary multiple-access interference only at the receiver and those underlay systems that, in addition, use cognition to adapt the secondary users' transmission parameters. We also noted that more work has been carried out for UWB underlay networks as opposed to DS-CDMA underlay networks.

We then presented an underlay cognitive system architecture for CDMA networks. We focused on improving the performance of the underlay system by minimizing the interference caused by the primary network under the condition that primary codes are not known to the secondary underlay network. To that end, we presented a novel technique that we developed in [35] for selecting the codes of the secondary users so that it minimizes the interference caused by the primary users. When compared to the performance of RCA in terms of the bit error rate significant improvement was demonstrated using IMCA. We also demonstrated several other measures indicating the robustness of IMCA in heavily loaded networks. In fact, the performance advantage of IMCA over RCA increased with increasing the number of users. The IMCA scheme is based on a primary CDMA network operating synchronously. We considered the asynchronous case in [36] and realized similar performance gains. It should be pointed out that the underlying concept behind IMCA is applicable to other primary networks as long as there is some periodicity in its transmission signal that may be exploited by judicious code assignment in the secondary network.

We then proposed a receiver architecture that applies the MOE technique as a natural candidate for secondary underlay networks. We also proposed a new technique to increase the convergence speed of the MOE by exploiting the knowledge of the codes of the secondary users.

Interweave cognitive radio may be quicker to the market due to its suitability for exploiting whitespaces in the TV band. In fact, the FCC has recently allowed devices to operate in the TV band provided they observe certain conditions [7]. These conditions include accessing a whitespaces database and spectrum sensing as well as following spectral masks for adjacent channel interference. Standards bodies have been developing standards, e.g. [71], for some time to utilize, and in anticipation, of freed spectrum after the analog to digital TV transition occurring in different parts of the world. There are still some challenges to overcome such as interference to wireless microphones. These are currently not required to register and hence do not appear in databases, are very difficult to sense, and are protected as incumbents. The high cooperation that overlay cognitive radio networks require from primary networks will probably limit their applicability except perhaps as secondary extensions of the same primary network. On the other hand, underlay cognitive radio networks have clear advantages in certain situations as outlined earlier and we believe they will find their place in the landscape of cognitive radio technologies.

References

[1] Mitola, J. (1999). "Cognitive radio for flexible mobile multimedia communications," in *Proc. Mobile Multimedia Commun.* (MoMuC '99), 3-10, Nov.

[2] Mitola, J. (2000). "Cognitive Radio: An Integrated Agent Architecture for Software Defined Radio," Ph.D. dissertation, KTH, Stockholm, Sweden, Dec. 2000.

[3] Srinivasa, S. & Jafar, S. A. (2007). "The throughput potential of Cognitive Radio: A theoretical perspective" *IEEE Communications Magazine*, May 2007, 74-79.

[4] P. J. Kolodzy, "Cognitive Radio Fundamentals," *SDR Forum*, Singapore, Apr. 2005.

[5] Federal Communications Commission, "Revision of part 15 of the commission's rules regarding ultra-wideband transmission systems, first report and order (ET Docket 98-153)," Adopted Feb. 14, 2002, Released Apr. 22, 2002.

[6] Ganesan, G. & Li, Y. G. (2005). "Cooperative spectrum sensing in cognitive radio networks," *Proc. IEEE DySPAN 2005*, November 2005, 137-143.

[7] Federal Communications Commission, "In the Matter of Unlicensed Operation in the TV Broadcast Bands, Additional Spectrum for Unlicensed Devices Below 900 MHz and in the 3 GHz Band (ET Docket 08-260)," Adopted Nov. 4, 2008, Released Nov. 14, 2008.

[8] Costa, M. H. M. (1983). "Writing on dirty paper," *IEEE Transactions on Information Theory*, vol. 29, 439-441, May 1983.

[9] Carleial, B. (1978). "Interference channels," *IEEE Transactions on Information Theory*, vol. 24, 60-70, Jan. 1978.

[10] Devroye, N., Mitran, P. & Tarokh, V. (2006). "Achievable Rates in Cognitive Radio Channels," *IEEE Transactions on Information Theory, vol. 52*, 1813-1827, May 2006.

[11] Jiang, J. & Xin, Y. (2007). "On the achievable rate regions for interference channels with degraded message sets," *IEEE Transactions on Information Theory,*vol. 54, issue 10,Oct. 2008.

[12] Mari´c, I., Goldsmith, A., Kramer, G. & Shamai, S. (2008). "On the capacity of interference channels with one cooperative transmitter," European Transactions on Telecommunications, vol. 19, issue 4, June 2008.

[13] Gel'fand, S. I. & Pinsker, M. S. (1980). "Coding for channel with random parameters," *Problemy Peredachi Informatsii*, vol. 9, no. 1, 19-31, 1980.

[14] Cover, T. M. & El Gamal, A. (1979). "Capacity Theorems for the Relay Channel," *IEEE Transactions on Information Theory*, vol. 25, 572-584, September 1979.

[15] Meulen, E. C. V. D. (1971). "Three-terminal communication channels," *Adv. Appl. Prob.*, vol. 3, 120-154, June 1971.

[16] Goldsmith, A., Jafar, S. A., Maric, I. & Srinivasa, S. (2009). "Breacking Spectrum Gridlock with Cognitive Radios: An Information Theoretic Perspective," *Proceedings of the IEEE*, vol. 97, Issue 5, 894-914, May 2009.

[17] Sahai, A., Hoven, N., Mishra, S. M. & Tandra, R. (2006). "Fundamental Tradeoffs in Robust Spectrum Sensing for Opportunistic Frequency Reuse," Technical Report, March 2006. Available online at http://www.eecs.berkeley.edu/»sahai/Papers/CognitiveTech-Report06.pdf.

[18] Tandra, R. & Sahai, A. (2008). "SNR Walls for signal detection," *IEEE Journal on Selected Topics in Signal Processing*, February 2008.

[19] Mishra, S. M., Sahai, A. & Brodersen, R. W. (2006). "Cooperative Sensing among Cognitive Radios," *IEEE International Conference on Communications*, vol. 4, 1658-1663, June 2006.

[20] Ghasemi, A. & Sousa, E. S. (2005). "Collaborative Spectrum Sensing for Opportunistic Access in Fading Environments," *IEEE International Symposium on New Frontiers in Dynamic Spectrum Access Networks*, 131-136, November 2005.

[21] Ganesan, G. & Li, Y. (2005). "Agility Improvement through Cooperative Diversity in Cognitive Radio," *IEEE Global Communications Conference, vol. 5*, November - December 2005.

[22] Tandra, R. & Sahai, A. (2007). "SNR Walls for Feature Detectors," IEEE International Symposium on New Frontiers in Dynamic Spectrum Access Networks, 559-570, April 2007.

[23] Sahai, A., Hoven, N. & Tandra, R. (2004). "Some fundamental limits in cognitive radio," *Allerton Conf. on Communication, Control and Computing* 2004, October 2004.

[24] Digham, F., Alouini, M. & Simon, M. (2003). "On the energy detection of unknown signals over fading channels," *Proc. IEEE ICC 2005*, vol. *5*, May 2003, 3575-3579.

[25] Ghasemi, A. & Sousa, E. S. (2005). "Collaborative spectrum sensing for opportunistic access in fading environment," *Proc. IEEE DySPAN 2005*, November 2005, 131-136.

[26] Cabric, D., Mishra, S. M. & Brodersen, R. W. (2004). "Implementation issues in spectrum sensing for cognitive radios," *Proc. 38th Asilomar Conference on Signals*, Systems and Computers 2004, November 2004, 772-776.

[27] Fehske, A., Gaeddert, J. D. & Reed, J. H. (2005). "A new approach to signal classification using spectral correlation and neural networks," *Proc. IEEE DySPAN 2005*, November 2005, 144-150.

[28] Tang, H. (2005). "Some physical layer issues of wide-band cognitive radio system," *Proc. IEEE DySPAN 2005*, November 2005, 151-159.

[29] Ghasemi, A. & Sousa, E. S. (2006). "Capacity of Fading Channels Under Spectrum-Sharing Constraints," *IEEE International Conference on Communications*, June 2006.

[30] Telatar, E. (1999). "Capacity of Multi-antenna Gaussian Channels," *European Transactions on Telecommunications*, vol. 10, 585-596, November 1999.

[31] Zhang, R. & Liang, Y. C. (2007). "Exploiting Multi-Antennas for Opportunistic Spectrum Sharing in Cognitive Radio Networks," Preprint available at http://arxiv.org/abs/0711.4414v1, November 2007.

[32] Chiani, M., Giorgetti, A. & Liva, G. (2005). "Ultra Wide Bandwidth communications towards Cognitive Radio, "EMC Europe Workshop 2005, Rome, Italy, 114-117.

[33] Arslan, H. & Sahin, M. E. (2007). UWB Cognitive Radio, ch. Cognitive Rdaio, Software Defined Radio, and Adaptive Wireless Systems. Springer.

[34] Sahin, M. E. & Arslan, H. (2006). "System design for cognitive radio communications," in *Proc. Cognitive Radio Oriented Wireless Networks and Commun. (CrownCom)*, (Mykonos Island, Greece), June 2006.

[35] Elezabi, A., Kashef, M., Abdallah, M. & Khairy, M. "CDMA Underlay Network with Cognitive Interference-Minimizing Code Assignment and Semi-Blind Interference Suppression," *Journal on Wireless Communication and Mobile Computing (Special Issue: Cognitive Radio and Advanced Spectrum Management),* vol. 9, issue 11, November 2009, DOI:10.1002/ wcm.724.

[36] Elezabi, A., Kashef, M., Abdallah, M. & Khairy, M. (2009). "Cognitive Interference-Minimizing Code Assignment for Underlay CDMA Networks in Asynchronous Multipath Fading Channels," *Proceedings of IWCMC*, Leipzig, June 2009.

[37] Lee, L. B. & Hossain, E. (2008). "Rcsource Allocation for Spectrum Underlay in Cognitive Radio Networks," *IEEE Transactions on Wireless Communications*, vol. 7, no. 12, 5306-5315, Dec. 2008.

[38] Laster, J. D. & Reed, J. H. (1997). "Interference Rejection in Digital Wireless Communications," *IEEE Signal Processing Magazine*, vol. 14, no. 3, 37-62, May 1997.

[39] Proakis, J. G. (1996). "Interference Suppression in Spread Spectrum Systems," *IEEE 4th International Symposium on Spread Spectrum Technology and Applications*, vol. 1, 259-266.

[40] Papoulis, and S. Unnikrishna Pillai, Probability, Random Variables, and Stochastic Processes. Fourth Edition, Boston, MA: McGraw Hill, 2002.

[41] Ketchum, J. W. & Proakis, J. G. (1982). "Adaptive Algorithms for Estimating and Suppressing Narrow-Band Interference in PN Spread-Spectrum Systems," *IEEE Transactions on Communications*, vol. 30, no. 5, 913-924, May 1982.

[42] Milstein, L. B. (1988). "Interference Rejection Techniques in Spread Spectrum Communications," *Proceedings of the IEEE*, vol. 76, no. 6, 657-671, June 1988.

[43] Iltis, R. A. & Milstein, L. B. (1984). "Performance Analysis of Narrow-Band Interference Rejection Techniques in DS Spread-Spectrum Systems," *IEEE Transactions on Communications*, vol. 32, no. 11, 1169-1177, Nov. 1984.

[44] Vijayan, R. & Poor, H. V. (1990). "Nonlinear Techniques for Interference Suppression in Spread Spectrum Systems," *IEEE Transactions on Communications*, vol. 38, no. 7, 1060-1065, July 1990.

[45] Milstein, L. B. & Das, P. K. (1977). "Spread Spectrum Receiver Using Surface Acoustic Wave Technology," *IEEE Transactions on Communications*, vol. 25, no. 8, 841-847, Aug. 1977.

[46] Milstein, L. B. & Das, P. K. (1980) "An Analysis of a Real-Time Transform Domain Filtering Digital Communication System – Part I: Narrow-Band Interference Rejection," *IEEE Transactions on Communications*, vol. 18, no. 6, 816-824, June 1980.

[47] Huang, X. & Liang, D. (1999). "Gradual RELAX algorithm for RFI suppression in UWB-SAR," *Electronic Letters*, Vol. 35, no. 22, 1916-1917, Oct. 1999.

[48] Juhel, B., Vezzosi, G. & Le Goff, M. (1999). "Radio Frequency Interference Suppression for Noisy Ultra Wide Band SAR Measurements," *Ultra-Wideband Short-Pulse Electromagnetics*, 387-393, New York: Kluwer Academic/Pienum Publishers, 1999.

[49] Ohno, K. & Ikegami, T. (2007). "Interference DAA technique for coexisting UWB radio," in *Proceedings of the 65th Vehicular Technology Conference* (VTC '07), 2910-2914, Dublin, Ireland, April 2007.

[50] Sum, S., Rahman, M. A., Sasaki, S., Harada, H. & Kato, S. (2008). "Cognitive Interference-Mitigation Technique for Hybrid DS-Multiband-UWB Multiple Access System," *2008 IEEE Radio and Wireless Symposium,* 105-108, Jan 2008.

[51] Chiang, J. & Lansford, J. (2005). "Use Of Cognitive Radio Techniques for OFDM UltraWideBand Coexistence with Wimax," *Texas Wireless Symposium*, 91-95.

[52] Paquelet, S., Froc, G., Bisiaux, A. & Mallégol, S. (2007). "Ultrawide Band: Issues and perspectives for an unlicensed radio," *The 13th European Wireless Conference,* Paris, Apr 2007.

[53] Giancola, G., Domenicali, D. & Di Benedetto, M. (2006). "Cognitive UWB: interference mitigation by spectral control," *1st International Conference on Cognitive Radio Oriented Wireless Networks and Communications*, 1-6, Jun 2006.

[54] Zhang, H. & Kohno, R. (2003). "Re-configurable soft-spectrum UWB

receiving scheme with multi-mode and multi-rate adaptation," in *Proceedings of the 6th International Symposium on Wireless Personal Multimedia Communications* (WPMC'03), Yokosuka, Japan, October 2003.

[55] Zhang, H., Zhou, X., Yazdandoost, K. Y. & Chlamtac, I. (2006). "Multiple signal waveforms adaptation in cognitive ultrawideband radio evolution," *IEEE Journal on Selected Areas in Communications*, vol. 24, no. 4, 878-884, 2006.

[56] Ohno, K. & Ikegami, T. (2006). "Interference mitigation study for UWB radio using template waveform processing," *IEEE Transactions on Microwave Theory and Techniques*, vol. 54, no. 4, 1782-1792.

[57] Ohno, K. & Ikegami, T. (2006). "Interference detection and mitigation by using multi-carrier template wave for pulse based UWB," in *Proceedings of the 9th IEEE International Symposium on Spread Spectrum Techniques and Application (ISSSTA '06)*, 183-187, Manaus, Brazil, August 2006.

[58] da Silva & Milstein, L. B. (2003). "Spectral-Encoded UWB Communication Systems," *2003 IEEE Conference on Ultra-Wideband Systems and Technologies*, 96-100, Nov. 2003.

[59] Elezabi, A., Kashef, M., Abdallah, M. & Khairy, M., "Interference-Minimizing Code Assignment Technique for Underlay Cognitive CDMA Network", US Patent Application 61291349 *(filed Dec. 30, 2009)*

[60] Kashef, M., Abdallah, M., Elezabi, A., & Khairy, M., ""System Parameter Selection for Asymmetric Underlay CDMA Networks with Interference-Minimizing Code Assignment", *Proc. of IEEE SPAWC June 2009, Perugia, Italy.*

[61] Kafle, P. L. & Sesay, A. B. (2004). "Iterative Semi-Blind Multiuser Detection for Coded MC-CDMA Uplink System," *IEEE Trans. Communications*, Vol. 51, No. 7, 1034-1039, March 2004.

[62] Høst-Madsen, A. & Cho., K. S. (1999). "MMSE/PIC Multiuser Detection for DS/CDMA Systems with Inter- and Intra-Cell Interference." *IEEE Trans. Communications*, Vol. 47, No. 2, 291-299, Feb. 1999.

[63] Mao, Z. & Bhargava, V. K. (2005). "Semiblind Linear Parallel Interference Cancellation Multiuser Detectors for DS-CDMA Systems, " *IEEE Trans. Wireless Communications*, VOL. 4, NO. 4, 1319-1324, July 2005.

[64] Honig, M., Madhow, U. & Verdu, S. (1995). "Blind adaptive multiuser detection," *IEEE Trans. Inform Theory*, vol. 41, 944-960, July 1995.

[65] Lupas, R. & Verdu, S. (1989). "Linear Multiuser Detectors for synchronous Code-Division Multiple Access Channels, " *IEEE Trans. Inform. Theory*, Vol. IT-35, 123-136, January 1989.

[66] Varanasi, M. K. & Aazhang, B. (1990). "Multistage detection in

asynchronous code-division multiple-access communications", *IEEE Trans. Comm.*, vol. 38, 509-519, Apr. 1990.

[67] Schober, R., Gerstacker, W. H. & Lampe, L. (2004). "Data-aided and blind stochastic gradient algorithms for widely linear MMSE MAI suppression for DS-CDMA," *IEEE Trans. Signal Processing*, Vol. 52, No. 3, 746-756, March 2004.

[68] Wolkowicz, H., Saigal, R. & Vandenberghe, L. (2000). *"Handbook on Semidefinite Programming"*, Kluwer.

[69] Laneman, J. N., Tse, D. N. C. & Wornell, G. W. (2004). "Cooperative diversity in wireless networks: Efficient protocols and ourage behavior", *IEEE Transaction on Information Theory*, vol.51, no.12, Dec. 2004.

[70] Boyer, J., Falconer, D. & Yanikomeroglu, H. (2001). "A theoretical characterization of the multihop wireless communications channel with diversity", in *Proc. IEEE GLOBECOM'01*, San Antonio, Texas, USA, November 2001.

[71] IEEE 802.22 Working Group on WRANs, http://www.ieee802.org/22

In: Cognitive Radio
Editors: P.D. McGuire et al, pp. 37-72

ISBN: 978-1-60876-604-8

Chapter 2

EVOLUTIONARY ALGORITHMS FOR COGNITIVE RADIO ADAPTATION

Shilian Zheng and Caiyi Lou
JiangNan Electronic Communication Research Institute,
Jiaxing 314033, China

Abstract

A primary feature of cognitive radios is the ability to adapt the transmission parameters given a dynamic wireless environment. In this chapter, the cognitive radio adaptation is formalized as a multi-objective optimization problem and four evolutionary algorithms, i.e., genetic algorithms (GAs), quantum genetic algorithm (QGA), particle swarm optimization (PSO), and shuffled frog leaping algorithm (SFLA), are explored and used to intelligently solve this problem and determine the optimal parameters for a given situation.

GAs have been regarded as an efficient algorithm for cognitive radio adaptation. However, classic GAs are quite poor at hill climbing and they may converge to some local maximum. In order to improve the convergence performance of the adaptation process, we have proposed to use QGA and PSO for the adaptation method in our previous work. This chapter reviews these three methods and compares their performance. After overview of GAs, QGA and PSO, we focus on a newly developed evolutionary algorithm, SFLA, which has not been investigated as thoroughly as GAs or PSO. We propose a binary SFLA in this chapter which can be used to solve any combinatorial optimization problem in general, and the proposed binary SFLA is used as the adaptation method for cognitive radios. After investigation binary SFLA-based cognitive radio adaptation, we present some challenges and future directions.

1. Introduction

Cognitive radio is a radio that can sense, learn and adapt to the surrounding environment according to its inner and outer stimuli [1-3]. The goals of adaptation include maximizing spectral efficiency, maximizing data throughput, minimizing interference to other cognitive radios, maximizing battery energy efficiency, etc. [4] Research done at Virginia Tech has developed a cognitive engine to perform the sensing, learning and adaptation tasks of a cognitive radio [5,6]. This chapter focuses solely on the adaptation mechanism, i.e., how to adjust the radio parameters to optimize the desired objectives.

Rondeau et al. [5,7] discussed the adaptation mechanism which uses a multi-objective genetic algorithm (GA) in detail. Their hardware experiments and software simulations validate that their GA implementation does in fact change the radio parameters to different settings based on a set of objectives. Hauris [8] also presented relations between fitness measures and radio frequency (RF) parameters and used a GA to adapt the RF parameters. Newman et al. [9] extended the adaptation to multicarrier systems, transformed multi-objectives into a signal objective via a simple weighted sum approach, and used GAs for optimization. Baynast et al. [10] invoked GAs for ARQ-based cross-layer optimization on cognitive radio networks. Their approach is more higher-layers and cooperative-networking based compared to the previous work [6-9]. In order to improve the performance of the GA-based adaptation method, we have combined simulated annealing with GA for cognitive radio optimization to improve GA's ability of hill-climbing at the later stage of the search [11]. We've also used a quantum GA (QGA) for cognitive radio adaptation to improve the convergence rate of the standard GA [12]. Newman et al. [13] proposed to seed the initial generation of the GA with chromosomes from the final generation of the previous GA cycle to speed up the convergence, thus decreasing the amount of time required to reach an optimal decision for a GA-based cognitive engine. In [14], we invoked another emerging population-based optimization method, particle swarm optimization (PSO), as an adaptation method for cognitive radios. In this chapter, we will review classic GAs, QGA and PSO for cognitive radio adaptation.

GAs, QGA (QGA can also be treated as a special case of GA) and PSO are three examples of evolutionary algorithms, which are stochastic search methods that mimic the metaphor of natural biological evolution and/or the social behavior of species [15]. Shuffled frog leaping algorithm (SFLA) [16] is a newly developed evolutionary algorithm which has not been investigated as thoroughly as GAs or PSO. In order to make SFLA more suitable for solving combinatorial optimization problem, we will propose a binary SFLA in this chapter and the proposed binary SFLA is used as the adaptation method for cognitive radios.

The rest of the chapter is organized as follows. In Section 2, the problem of cognitive radio adaptation is formalized. In section 3, 4 and 5, we reviewed the adaptation method based on GAs, QGA and PSO, respectively. Performance comparison of these three methods is presented in Section 5. Section 6 focuses on the proposed binary SFLA and the adaptation method based on the proposed binary SFLA. Section 7 presents some challenges and future notes. Finally, conclusion is provided in Section 8.

2. Adaptation in Cognitive Radios

Cognitive radio adaptation and optimization must occur in many dimensions including user domain, radio hardware domain, radio frequency (RF) domain, and policy domain, and is subject to a variety of constraints [7]. A general formulation of cognitive radio adaptation problem is presented in the following subsections.

2.1. Parameters and Objectives

The parameters of a radio are any of the variables that affect link performance and radio operation. Environment variables inform the system of the characteristics of the surrounding environment, while decision variables are under real-time control of either the operator or the radio's real-time control processes. Cognitive radios use the control parameters made available by the underlying software defined radios [13]. On the physical (PHY) layer, carrier frequency, transmit power, modulation type and index, symbol rate can all be adjusted. On the medium access control (MAC) layer, adjustable parameters may include channel coding rate, frame size, and multiple access technique.

The general purpose of a cognitive radio is to autonomously strive for improving the system performance. Important performance measurements include bit error rate (BER), data throughput, and actual radio power consumption. Each metric has unique relationships with the other metrics, and different parameters alter different metrics in different ways [2]. For instance, in order to increase the data rate, we could use a higher modulation index, but this increases the BER as well. Decreasing the symbol rate will decrease the BER, but at the expense of the data rate. To facilitate the decision making in a cognitive engine, each performance objective should be represented by a mathematical relationship that relates a set of parameters to a scalar value describing how well this set achieves the specific goal.

2.2. Constraints

While a cognitive radio tries to optimize on the PHY and MAC layers to maximize the performances, it must respect to any local regulatory restricts. Devices operating in different frequency bands will be subject to different constraints. Certain bands would have strong restrictions against certain power level transmission (GPS), locations (TV broadcast), or time (public safety) [7]. Besides regulatory limitations, the hardware also provides constraints for decisions that cognitive engine can make. The cognitive radio should also be aware of the constraints imposed by computational power and battery power available for a given waveform. A system with limited power might find a narrowband modulation more suitable than orthogonal frequency division multiplexing (OFDM) waveform.

2.3. Preferences

Different users, applications, and environments may lead to different preferences to each objective. The user who is to transmit large amount of digital files cares more about BER and data rate, while a video conference user cares more about network latency and picture jitter. Transferring an email does not require high data rate while broadband video needs high throughput. In RF environment with strong noise and interference, BER should be paid more attention to, while in a mild situation where little interference exists, spectral efficiency may be the main concern. Adaptation in cognitive radios should proceed intelligently such that the primary objectives are mainly optimized while secondary objectives are not overlooked.

2.4. Multi-objective Optimization

The basic process followed by a cognitive radio is that it adjusts its parameters to achieve some desired combination of objectives. We denote the adjustable parameters as $\boldsymbol{a} = [a_1, a_2, ..., a_n]$ and objective functions which a cognitive radio needs to optimize as $\boldsymbol{f} = [f_1, f_2, ..., f_m]$, where n is the number of control parameters and m is the number of objective functions. In this framework, the cognitive radio adaptation process becomes a multi-objective optimization problem, which can be represented as:

$$\begin{aligned} &\min/\max\{\boldsymbol{f}(\boldsymbol{a})\} = [f_1(\boldsymbol{a}), f_2(\boldsymbol{a}), ..., f_m(\boldsymbol{a})] \\ &\text{subject to} : \boldsymbol{a} \in A, \boldsymbol{f} \in F, \end{aligned} \tag{1}$$

where A is the parameter space and F the objective space.

To represent the relative importance of the objective functions in the decision-making process, we denote a weight vector as $\boldsymbol{w} = [w_1, w_2, ..., w_m]$, where w_i denotes the importance of the ith objective function and it satisfies the following conditions:

$$\begin{aligned} &w_i \geq 0, i = 1, 2, ..., m, \\ &w_1 + w_2 + ... + w_m = 1. \end{aligned} \tag{2}$$

To simplify the problem, we transform the multi-objective functions into a single objective function. Suppose all the objective functions in $\boldsymbol{f} = [f_1, f_2, ..., f_m]$ have been normalized to the range [0,1] and transformed to maximization problem, then the multi-objective functions can be transformed into the following objective function by a simple weighted sum approach [13, 17]:

$$f = \sum_{i=1}^{m} w_i f_i \, . \tag{3}$$

Consequently, the adaptation in a cognitive radio is to adjust parameters to maximize the objective function as shown in Equation (3).

It is worthwhile to note that once the weight vector is fixed, so is the optimization direction. This is the intended property when trying to find a single optimal solution for a given environment. Changing the objective direction requires only a simple change of the weight vector, which makes switching to a desired operational mode relatively easy.

3. Genetic Algorithms for Cognitive Radios

3.1. Introduction of GAs

GAs are rooted in the mechanisms of evolution and natural genetics. A solution to a given problem is represented in the form of a string called 'chromosome', consisting of 'genes' which hold a set of values for the optimization variables [18]. An encoding mechanism is used to map the optimization variables into genes of a chromosome. Binary encoding is considered in this chapter. Figure 1 summarizes the working of a simple GA [19].

As shown in Figure 1, after initialization, the fitness values of the solutions are evaluated and act as index to describe how well the chromosomes (solutions) to the problem are. Each iteration of the while-loop mimics one natural evolution. Selection models nature's survival-of-the-fittest mechanism and leads to fitter solutions survive while weaker ones perish. The roulette wheel selection scheme is a commonly used proportionate selection scheme. Chromosome i is allocated a sector of a roulette wheel with the angle subtended by the sector at the center of the wheel equaling $2\pi\rho_i / \rho_{avg}$ ($1 \le i \le P$), where ρ_i is the fitness value of chromosome i, ρ_{avg} is the average fitness value over the total population, and P is the population size. Crossover comes after selection. It is one of GA's crucial operations and mimics natural genetic recombination. Crossover exchanges portions of the randomly picked pair of strings if a randomly generated number in the range 0 to 1 is smaller than p_c, the crossover rate. After crossover, strings are subjected to mutation which simulates natural genetic variation. Mutation of a bit involves flipping it: changing a 0 to 1 and vice versa. The mutation rate p_m gives the probability that a bit will be flipped. The final step in each evolution cycle is the evaluation of fitness values which will be used for the selection in the next generation. To terminate execution of GA, a stopping criterion must be specified. If the maximum number of generations is reached, the algorithm is terminated.

The GA used in this study is steady state, which means only a fraction of strings but not the entire population are replaced by the new individuals [20]. In the steady-state GA, the best strings are preserved through the evolution, which is analogous to elitist strategy.

3.2. GAs-Based Cognitive Radio Adaptation

The first step to use classic GAs for cognitive radio adaptation is to map the radio parameters which need to be adjusted to genes in GAs. The structure of the chromosome is shown in Figure 2 for the adaptation problem in cognitive radios, where a_j will be replaced by a binary string after the encoding mechanism. The number of bits used to encode a parameter is determined by the range and the expected precision of the parameter. The wider the range and/or the higher the precision of the parameter, the more number of bits needed to encode it. For instance, encoding frequency which is several GHz wide and has a resolution 1 Hz may need more than 20 bits, while encoding modulation types may only need 3 or 4 bits.

```
Simple Genetic Algorithm()
{
   initialize population;
   evaluate population;
   while (termination conditions not reached)
   {
      select solutions for next population;
      perform crossover and mutation ;
      evaluate population;
   }
}
```

Figure 1. Simple genetic algorithm structure.

a_1	a_2	$\cdots$	a_n

Figure 2. The structure of the chromosome.

Fitness function is significantly important for GAs as it is almost the only outside information that GAs use in the evolution process. The objective function provides the mechanism for evaluating the fitness of each chromosome in the population. We directly use the objective function for fitness evaluation:

$$\rho = f = \sum_{i=1}^{m} w_i f_i . \qquad (4)$$

The reason why we do not use any scaling mechanism to enhance the probability that the best chromosome is chosen in the selection process is that we are using a steady-state GA, and the best chromosome is preserved through the evolution. The GA-based adaptation proceeds as follows:

Step 1: randomly generate an initial population.
Step 2: compute the fitness of each individual chromosome of the current population using Equation (4).
Step 3: select some chromosomes as an intermediate population using roulette wheel selection scheme.
Step 4: carry out the two-point crossover scheme.
Step 5: carry out the mutation operation.
Step 6: replace a fraction of the previous population, individuals which have lower fitness values, with new individuals.
Step 7: if it reaches the predefined maximum generation, then terminate the algorithm; if not, go to step 2.

Simulations and performance analysis of the algorithm will be presented in Section 5.3.

4. Quantum Genetic Algorithm for Cognitive Radios

4.1. Introduction of QGA

The classical representations can be broadly classified as binary, numeric and symbolic encoding in GAs, while QGA uses a novel representation that is based on the concept of qubits [21]. A qubit may be in state '1', in state '0', or in any superposition of the two. The state of a qubit can be represented as:

$$|\varphi\rangle = \alpha|0\rangle + \beta|1\rangle, \qquad (5)$$

where α and β are complex numbers that denote the probability amplitudes of state '0' and state '1' respectively. The act of observing a quantum state will make it collapse to a single state. $|\alpha|^2$ gives the probability that the qubit will be found in state '0', while $|\beta|^2$ gives the probability that the qubit will be found in state '1' [22]. α and β must satisfy the following condition:

$$|\alpha|^2 + |\beta|^2 = 1. \tag{6}$$

A chromosome with l-qubits can be represented as:

$$\boldsymbol{q} = \begin{bmatrix} \alpha_1 & \alpha_2 & \cdots & \alpha_l \\ \beta_1 & \beta_2 & \cdots & \beta_l \end{bmatrix}, \tag{7}$$

where $|\alpha_i|^2 + |\beta_i|^2 = 1, i = 1, 2, ..., l$. Note that QGA with qubit representation has a better characteristic of diversity than classical encoding mechanisms, as it can represent superposition of states. Convergence can also be obtained with the qubit representation [23]. As $|\alpha_i|^2$ approaches to 0 or 1, the qubit chromosome converges to a single state.

Using qubit representation, the population of QGA can be denoted as $\boldsymbol{Q}(g) = \{\boldsymbol{q}_1^g, \boldsymbol{q}_2^g, ..., \boldsymbol{q}_P^g\}$, where g is the generation, P denotes the population size, and $\boldsymbol{q}_i^g$ denotes the ith chromosome at the gth generation, which can be represented as:

$$\boldsymbol{q}_i^g = \begin{bmatrix} \alpha_{i1}^g & \alpha_{i2}^g & \cdots & \alpha_{il}^g \\ \beta_{i1}^g & \beta_{i2}^g & \cdots & \beta_{il}^g \end{bmatrix}. \tag{8}$$

α_{ij}^0 and β_{ij}^0 ($i = 1, 2, ..., P$, $j = 1, 2, ..., l$) are initialized to $1/\sqrt{2}$ on the initial population.

Observing every chromosome of $\boldsymbol{Q}(g)$ results in a set of states denoted as $\boldsymbol{O}(g)$, which is a set of binary solutions, where $\boldsymbol{O}(g) = \{\boldsymbol{o}_1^g, \boldsymbol{o}_2^g, ..., \boldsymbol{o}_P^g\}$, and $\boldsymbol{o}_i^g$ ($i = 1, 2, ...P$) is a binary string with length l. The value of the jth bit of $\boldsymbol{o}_i^g$ is determined by the jth qubit of $\boldsymbol{q}_i^g$. Randomly generate a number distributed in

[0,1], and if it is lower than $|\alpha_{ij}^g|^2$, then the *j*th bit of $\boldsymbol{o}_i^g$ takes '0'; otherwise, it takes '1'. Note that the fitness evaluation is carried out on $\boldsymbol{O}(g)$, but not on $\boldsymbol{Q}(g)$.

QGA employs quantum gates to update $\boldsymbol{Q}(g)$. We use the rotation gate in this chapter:

$$\begin{pmatrix} \alpha_{ij}^{'} \\ \beta_{ij}^{'} \end{pmatrix} = \begin{pmatrix} \cos\theta_{ij} & -\sin\theta_{ij} \\ \sin\theta_{ij} & \cos\theta_{ij} \end{pmatrix} \begin{pmatrix} \alpha_{ij} \\ \beta_{ij} \end{pmatrix}, \tag{9}$$

where $\begin{pmatrix} \alpha_{ij} \\ \beta_{ij} \end{pmatrix}$ and $\begin{pmatrix} \alpha_{ij}^{'} \\ \beta_{ij}^{'} \end{pmatrix}$ $(i = 1, 2, ..., P,\ \ j = 1, 2, ..., l)$ are the *j*th qubit of chromosome *i* before and after updating, respectively, and θ_{ij} denotes the rotation angle, the value of which is given as $\theta_{ij} = \Delta\theta_{ij} \cdot s(\alpha_{ij}, \beta_{ij})$, where $\Delta\theta_{ij}$ and $s(\alpha_{ij}, \beta_{ij})$ are shown in Table 1, where '*' means that the entry can take any real value, and '± 1' means the entry can take either 1 or -1 [24]. For instance, if the fitness value of $\boldsymbol{o}_i$, $\rho(\boldsymbol{o}_i)$, is lower than the fitness value of the best stored binary solution, $\rho(\boldsymbol{u})$, and o_{ij} and u_j are 0 and 1, respectively, then the value of θ_{ij} can be set as $+\text{Delta}$ or $-\text{Delta}$ according to the value of α_{ij} and β_{ij} to increase the probability of the state '1'.

4.2. QGA-Based Cognitive Radio Adaptation

The structure of the chromosome of QGA is similar to GAs as shown in Figure 2 except that a_j will be replaced by a qubit string rather than a binary string. Note that the observed binary string of a chromosome in QGA plays the same role as the binary encoded chromosome in GAs such that the observed state of a chromosome specifies a possible solution of the optimization problem. The number of bits used to encode a parameter is also determined by the range and the expected precision of the parameter.

QGA uses fitness function to evaluate the performance of each individual as in GAs. The difference is that QGA evaluates the fitness of the observed binary string but not the chromosome itself. As the chromosome is updated by the rotation gate which is directed by the fitness comparison of the observed binary string and the best stored solution as shown in Table 1, whether $\rho(\boldsymbol{o}_i)$ is larger than $\rho(\boldsymbol{u})$ significantly influences the update process, but the absolute value of $\rho(\boldsymbol{o}_i)-\rho(\boldsymbol{u})$ is not important at all. As such, we directly use the objective function for fitness evaluation without any scaling process. The QGA-based adaptation algorithm proceeds as follows:

Step 1: set $g=0$, and initialize $\boldsymbol{Q}(g)=\{\boldsymbol{q}_1^g,\boldsymbol{q}_2^g,...,\boldsymbol{q}_P^g\}$.

Step 2: observe every chromosome in $\boldsymbol{Q}(g)$ to obtain a set of states $\boldsymbol{O}(g)=\{\boldsymbol{o}_1^g,\boldsymbol{o}_2^g,...,\boldsymbol{o}_P^g\}$.

Step 3: compute the fitness values of $\boldsymbol{O}(g)$, store the best solution as $\boldsymbol{u}(g)$.

Step 4: if it reaches the predefined maximum generation, the algorithm is terminated, and $\boldsymbol{u}(g)$ is the best solution obtained by QGA for the adaptation problem; else, go to next step.

Step 5: set $g=g+1$.

Step 6: observe every chromosome in $\boldsymbol{Q}(g-1)$ to obtain a set of states $\boldsymbol{O}(g)=\{\boldsymbol{o}_1^g,\boldsymbol{o}_2^g,...,\boldsymbol{o}_P^g\}$.

Step 7: compute the fitness values of $\boldsymbol{O}(g)$.

Step 8: compare the fitness value of the best solution in $\boldsymbol{O}(g)$ and that of $\boldsymbol{u}(g-1)$, store the better one as $\boldsymbol{u}(g)$.

Step 9: employ quantum gates to update $\boldsymbol{Q}(g-1)$ using Equation (9) to obtain $\boldsymbol{Q}(g)$, and go to step 5.

Performance analysis of the algorithm will be presented in Section 5.3.

Table 1. Lookup table of the rotation angle.

o_{ij}	u_j	$\rho(\boldsymbol{o}_i) > \rho(\boldsymbol{u})$	$\alpha_{ij}\beta_{ij} > 0$		$\alpha_{ij}\beta_{ij} < 0$		$\alpha_{ij} = 0$		$\beta_{ij} = 0$	
			$s(\alpha_{ij},\beta_{ij})$	θ_{ij}	$s(\alpha_{ij},\beta_{ij})$	θ_{ij}	$s(\alpha_{ij},\beta_{ij})$	θ_{ij}	$s(\alpha_{ij},\beta_{ij})$	θ_{ij}
0	0	*	*	0	*	0	*	0	*	0
1	1	*	*	0	*	0	*	0	*	0
0	1	False	+1	+Delta	−1	−Delta	0	0	±1	±Delta
0	1	True	−1	−Delta	+1	+Delta	±1	±Delta	0	0
1	0	False	−1	−Delta	+1	+Delta	±1	±Delta	0	0
1	0	True	+1	+Delta	−1	−Delta	0	0	±1	±Delta

5. Particle Swarm Optimization for Cognitive Radios

5.1. Introduction of PSO

PSO was developed by Kennedy and Eberhart [25]. In PSO, each solution is a 'bird' in the flock and is referred to as a 'particle'. A particle is analogous to a chromosome in GAs [15]. However, the evolutionary process in PSO does not create new birds from parent ones. Rather, the birds only evolve their social behavior and accordingly their movement towards a destination [26].

The PSO developed in [25] is restricted in real number space. In order to solve optimization problems in a discrete number space, Kennedy and Eberhart [27] developed a discrete binary version of PSO, which is the focus of this chapter.

The following notation is needed in binary PSO. The number of particles in the swarm is denoted as S. Let $\boldsymbol{x}_i^t = [x_{i1}^t, x_{i2}^t, \ldots, x_{iD}^t]$ be the position of particle i ($1 \le i \le S$) at iteration t, where D is the number of dimensions to represent a particle and $x_{id}^t \in \{0,1\}$ is the dth ($1 \le d \le D$) dimension of the position of particle i. Note that $\boldsymbol{x}_i^t$ is analogous to the chromosome using binary encoding mechanism in GA, and it is treated as a potential solution of the optimization problem. The velocity of particle i at iteration t is denoted as $\boldsymbol{v}_i^t = [v_{i1}^t, v_{i2}^t, \ldots, v_{iD}^t], \boldsymbol{v}_{id}^t \in R$. Each particle in the swarm is assigned a fitness value indicating the merit of this particle such that the swarm evolution is navigated by best solutions. Let $\boldsymbol{p}_i^t = [p_{i1}^t, p_{i2}^t, \ldots, p_{iD}^t]$ be the best solution that particle i has obtained until iteration t, and $\boldsymbol{p}_b^t = [p_{b1}^t, p_{b2}^t, \ldots, p_{bD}^t]$ be the best solution obtained from $\boldsymbol{p}_i^t$ in the population at iteration t.

Each particle adjusts its velocity according to the previous velocity, the cognition part and the social part [28]. The velocity of the dth dimension of the ith particle is updated as follows:

$$v_{id}^t = \omega v_{id}^{t-1} + c_1 r_1 (p_{id}^{t-1} - x_{id}^{t-1}) + c_2 r_2 (p_{bd}^{t-1} - x_{id}^{t-1}), \qquad (10)$$

where c_1 and c_2 are two positive constants named learning factors or acceleration coefficients, r_1 and r_2 are uniform random numbers distributed in the range [0,1], and ω is an inertia weight employed to control the impact of the previous history of velocities on the current velocity [26].

For the velocity value of each bit in a particle, Kennedy and Eberhart [27] claim that higher value is more likely to choose 1, while lower value favors the 0 choice. Furthermore, they constrain the velocity value to the range [0, 1] by using the following sigmoid function:

$$\text{sig}(v_{id}^t) = \frac{1}{1+\exp(-v_{id}^t)}, \tag{11}$$

where $\text{sig}(v_{id}^t)$ denotes the probability of x_{id}^t taking 1. If $r < \text{sig}(v_{id}^t)$, then $x_{id}^t = 1$; else, $x_{id}^t = 0$, where r is a random number uniformly distributed in [0,1]. The continuous-valued particle swarm algorithm always limit v_{id}^t by a value $V_{\max}$ which is a parameter of the system. In the discrete version $V_{\max}$ is retained to avoid $\text{sig}(v_{id}^t)$ approaching 0 or 1, i.e., $v_{id}^t \in [-V_{\max}, +V_{\max}]$ [29].

5.2. PSO-Based Cognitive Radio Adaptation

As in GAs and QGA, the first step to use binary PSO for cognitive radio adaptation is to map the radio parameters which need to be adjusted to position of particles in binary PSO. The structure of position $\boldsymbol{x}_i^t = (x_{i1}^t, x_{i2}^t, \ldots, x_{iD}^t)$ is shown in Figure 3 for the adaptation problem in cognitive radios, where L_j denotes the number of bits to encode parameter a_j $(1 \le j \le n)$. The number of bits used to encode a parameter is determined by the range and the expected precision of the parameter.

The objective function provides the mechanism for evaluating the fitness of each particle. As the goal of cognitive radio adaptation is to maximize the objective function represented by Equation (3), we directly use the objective value computed from Equation (3) as the fitness value of the corresponding particle. The adaptation follows the procedure of the binary PSO, and when the binary PSO is terminated, the position of the global best particle $\boldsymbol{p}_b^t = [x_{b1}^t, x_{b2}^t, \ldots, x_{bD}^t]$ is the best solution obtained which will be used to configure the cognitive radio.

In conclusion, the PSO-based cognitive radio adaptation and optimization algorithm proceeds as follows:

Step 1: set $t=0$, and randomly generate x_{id}^t and v_{id}^t, where $x_{id}^t \in \{0,1\}$, $v_{id}^t \in [-V_{\max}, +V_{\max}]$, $1 \le d \le D$, thus obtaining $x_i^t = [x_{i1}^t, x_{i2}^t, \ldots, x_{iD}^t]$, $1 \le i \le S$.

Step 2: compute the fitness value of each particle in the population and set $\boldsymbol{p}_i^t = [x_{i1}^t, x_{i2}^t, \ldots, x_{iD}^t]$ and $\boldsymbol{p}_b^t = [x_{b1}^t, x_{b2}^t, \ldots, x_{bD}^t]$, where *b* is the index of the particle which has the highest fitness value.

Step 3: set $t=t+1$, and update v_{id}^t according to Equation (6). If $v_{id}^t > V_{\max}$, then set $v_{id}^t = V_{\max}$; if $v_{id}^t < -V_{\max}$, set $v_{id}^t = -V_{\max}$.

Step 4: randomly generate a real number uniformly distributed in [0,1], if it is smaller than $\mathrm{sig}(v_{id}^t)$, then $x_{id}^t = 1$; else, $x_{id}^t = 0$.

Step 5: compute the fitness value of each particle in the population. For particle *i*, if it's fitness value is greater than the fitness value of $\boldsymbol{p}_i^{t-1}$, then set $\boldsymbol{p}_i^t = [x_{i1}^t, x_{i2}^t, \ldots, x_{iD}^t]$. If particle *i*'s fitness value is greater than the fitness value of $\boldsymbol{p}_b^{t-1}$, then set $\boldsymbol{p}_b^t = [x_{i1}^t, x_{i2}^t, \ldots, x_{iD}^t]$.

Step 6: if *t* equals to the predefined maximum iteration, then the algorithm is terminated; otherwise, go to step 3.

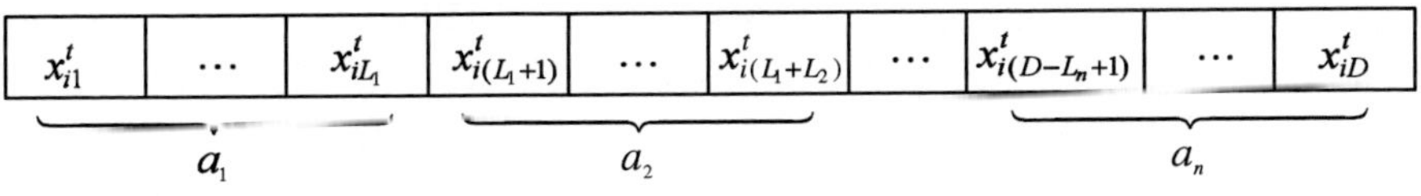

Figure 3. The structure of the position of particle *i* at iteration *t*.

5.3. Simulation Analysis

5.3.1. Multicarrier System Design

We simulated a multicarrier system with 32 subcarriers similar to [9]. The noise floor is assumed to be -90 dBm and a constant path loss from the transmitter to the receiver is assumed to be 78 dB. In addition, each subcarrier is assigned a random attenuation value which ranges from 0 to 1 to simulate a dynamic channel. The transmit power ranges from 0 dBm to 25.2 dBm with an increment of 0.4 dBm. The modulation types considered in the simulation are BPSK, QPSK, 16QAM and 64QAM. In this case the total search space is $(64\times 4)^{32} \approx 1.158\times 10^{77}$ and it needs 256 bits to represent a potential solution. Three objective functions are used which are minimize transmit power, minimize BER and maximize data rate. They are defined as follows:

$$f_{\text{min\_power}} = 1 - \frac{\frac{1}{N}\sum_{i=1}^{N} P_i}{P_{\max}}, \tag{12}$$

$$f_{\text{min\_ber}} = \left(1 - \frac{\log_{10}(0.5)}{\log_{10}(\overline{P}_{be})}\right)^4, \tag{13}$$

$$f_{\text{max\_data\_rate}} = \frac{\frac{1}{N}\sum_{i=1}^{N}\log_2 M_i - \log_2 M_{\min}}{\log_2 M_{\max} - \log_2 M_{\min}}, \tag{14}$$

where P_i is the transmit power over the ith subcarrier, $P_{\max}$ is the maximum available transmit power, $\overline{P}_{be}$ is the average BER over N channels, $M_{\min}$ is the minimum modulation order, $M_{\max}$ is the maximum modulation order, M_i is the modulation order used in the ith subchannel, and N is the number of subcarriers (channels). Assuming a gray-coded bit assignment and an AWGN channel model, the BER can be computed using Equations in [30]. The three objective functions are combined as follows:

$$f = w_1 f_{\text{min\_power}} + w_2 f_{\text{min\_ber}} + w_3 f_{\text{max\_data\_rate}}\,. \tag{15}$$

In this case, the adaptation mechanism needs to adjust the parameters to maximize Equation (15). Note that BER calculation involves computation of the complementary error function $\text{erfc}(x)$ which can be approximated by the following series to decrease the computation complexity [2]:

$$\operatorname{erfc}(x)=\begin{cases}1-\frac{1}{\sqrt{\pi}}\left(x-\frac{x^3}{3}+\frac{x^5}{10}+\cdots\right), & \text{for } x<3 \text{ with 40 items in the series,}\\ \frac{e^{-x^2}}{x\sqrt{\pi}}\left(1-\frac{1}{2x^2}+\frac{1\cdot 3}{2^2x^4}-\frac{1\cdot 3\cdot 5}{2^3x^6}+\cdots\right), & \text{for } x\geq 3 \text{ with 10 items in the series.}\end{cases} \tag{16}$$

Four different weight settings shown in Table 2 are used to test the effectiveness of the proposed method. Each mode emphasizes different objectives and is appropriate for different situations and applications. Mode 1 emphasizes on minimizing power and it is for applications like text messaging. Mode 2 is for emergency applications and mode 3 is for applications requiring high throughput like broadband video, while mode 4 has no preference to any specific objective.

5.3.2. Control Parameter Setting for GA, QGA and PSO

Each algorithm has its own control parameters that affect its performance in terms of processing time and solution quality. The control parameters of GA, QGA and PSO are listed in Table 3.

In the simulations, we set the control parameters such that the total times of fitness evaluation is the same in the three algorithms. For GA, the population size is set to 30, and the crossover probability and the mutation probability are set to 0.6 and 0.001, respectively. The evolutionary process is terminated after 1500 generations. The GA is configured to replace 80% of its population each generation, 24 of every 30 population members. As for QGA, the population size and the maximum generation are set the same as GA, 30 and 1500, respectively. Delta is decreased linearly from 0.1π at the first generation to 0.005π at the last generation. The parameters for the PSO are $S=30$, $c_1=c_2=2$, and $V_{\max}=4$, and PSO would be terminated after 1500 iterations.

Table 2. Weight settings.

Weights	Mode 1	Mode 2	Mode 3	Mode 4
w_1	0.80	0.15	0.05	1/3
w_2	0.15	0.80	0.15	1/3
w_3	0.05	0.05	0.80	1/3

Table 3. Control parameters of GA, QGA and PSO.

Algorithms	Control parameters	Comments
GA	Population size	Low population size may accelerate the convergence of the algorithm, but the population lacks diversity and the algorithm may result in premature convergence. Large population size may need long processing time. The suggested population size is between 20 and 100.
	Crossover rate	Crossover is a dominant process in GA to generate new individuals. As such, crossover rate is always set to a high value. However, if crossover rate is too high, the excellent pattern may be destroyed, which is not beneficial for the evolution. The suggested crossover rate is between 0.4 and 0.99.
	Mutation rate	Mutation can generate new individuals that cannot be generated by crossover. If mutation rate is too low, the ability to restrain premature convergence is weakened. However, if mutation rate is too high, GA may perform as a random search method. The suggested mutation rate is between 0.0001 and 0.1.
	Maximum generation	Maximum generation is highly related to the complexity of the optimization problem. In general, the larger the maximum generation, the better the solution. The suggested maximum generation is between 100 and 2000.
QGA	Population size	Similar to population size in GA
	Maximum generation	Similar to maximum generation in GA
	Delta	High Delta may accelerate the convergence of the algorithm, but the algorithm may converge prematurely. If Delta is too low, the algorithm will converge slowly. The suggested Delta is between 0.001π and 0.2π.

Table 3. Continued

Algorithms	Control parameters	Comments
PSO	Swarm size	Similar to population size in GA
	Inertia weight	High inertia weight is beneficial for global search, while low inertia weight strengthens local search. In the basic PSO, the inertia weight is set to 1.
	Learning factors	The two learning factors specify the abilities to push the particle to its own best position and the global best position of the entire swarm, respectively. The suggested learning factors are equal to 2.
	Maximum velocity	Maximum velocity specifies the allowed maximum distance the particle is moving at each iteration. it is also a parameter which is a tradeoff between the global search ability and the local search ability. The commonly used maximum velocity is 4.
	Maximum number of iterations	Similar to maximum generation in GA

5.3.3. Results Discussion

Ten independent experiments are conducted to each mode using GA, QGA, and PSO respectively. The objective function value of the chromosome or particle which has the highest fitness value at each iteration is recorded and the resulting average objective function values over 10 simulation runs are shown in Figure 4. The exact average objective function values after 50, 100, 500 and 1500 iterations are shown in Table 4. Figure 4 and Table 4 illustrate that QGA and PSO perform much the same as each other, QGA and PSO converges faster than GA, and the solutions optimized by the QGA or PSO algorithm have far higher objective values than those optimized by the GA. The standard deviations of the objective values after 1500 iterations over 10 runs are also given in Table 4, and it can be seen that the PSO and the QGA are more stable than the GA.

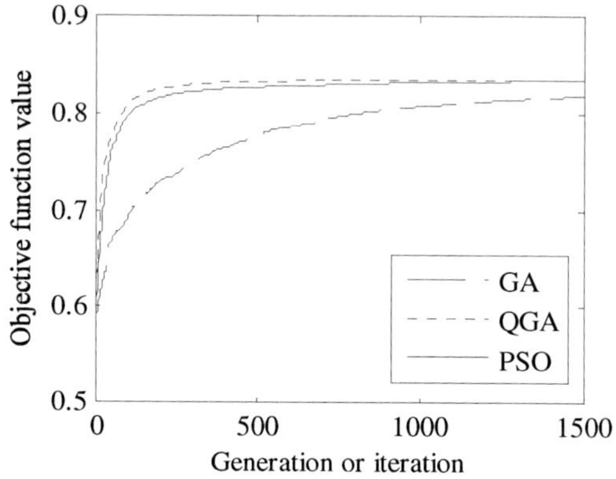

(a)

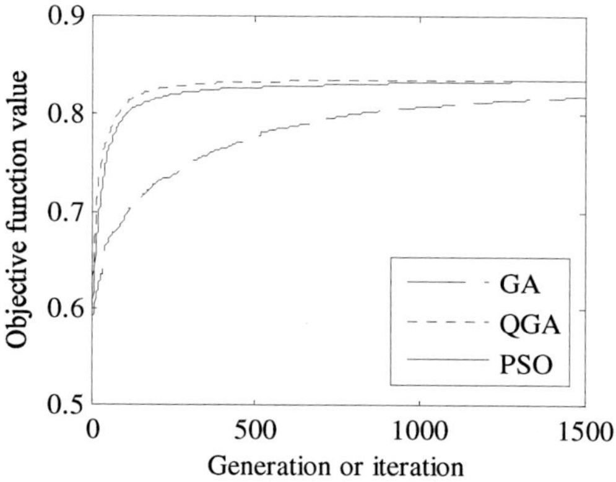

(b)

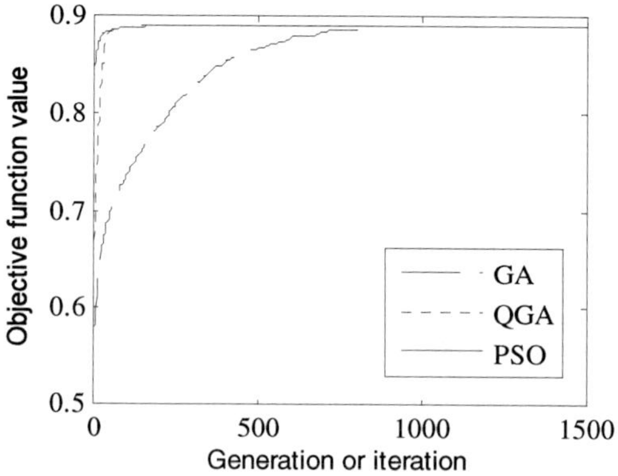

(c)

Figure 4. Continued on next page.

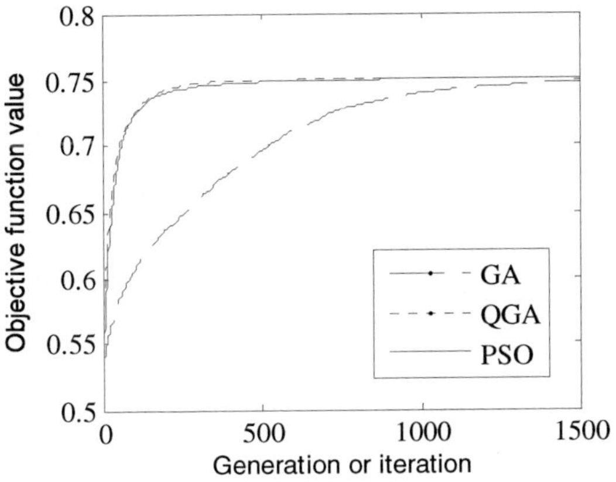

(d)

Figure 4. Performance comparison: GA versus QGA versus PSO. (a) mode 1; (b) mode 2; (c) mode 3; (d) mode 4.

6. Shuffled Frog Leaping Algorithm for Cognitive Radios

6.1. Introduction of SFLA

Traditional evolutionary algorithms rely on the concept of population. A population is a set of individuals. Each individual has a fitness value that evaluates the goodness of the individual [31]. In the SFLA, the population consists of a set of frogs (solutions) that is partitioned into subsets referred to as memeplexes that are permitted to evolve independently to search the space in different directions. Within each memeplex, the individual frogs hold ideas, that can be infected by the ideas of other frogs, and evolve through a process of memetic evolution [15]. Memetic evolution improves the quality of the meme of an individual and enhances the individual's performance towards a goal. After a certain number of memetic evolution steps, the memeplexes are foced to mix and new memeplexes are formed through a shuffling process. The shuffling enhances the quality of the memes after being infected by frogs from different regions of the swamp. The local search and the shuffling processes continue until defined convergence criteria are satisfied [16].

For D-dimensional problems, a frog i is represented as $\boldsymbol{y}_i = (y_{i1}, y_{i2}, \ldots, y_{iD})$, $1 \le i \le P$, where P is the population size. After randomly generate the initial population, the frogs are sorted in a descending order according to their fitness values. Then, the entire population is divided into k memeplexes, each of which containing ℓ frogs. The first frog goes to the first memeplex, the second frog goes to the second memeplex, the kth frog goes to the kth memeplex, and frog k+1 goes back to the first memeplex, and so on [15].

In each memeplex, the frog with the best fitness is denoted as $\boldsymbol{y}_b = (y_{b1}, y_{b2}, ..., y_{bD})$, and the frog with the worst fitness $\boldsymbol{y}_w = (y_{w1}, y_{w2}, ..., y_{wD})$. The frog with the global best fitness is identified as $\boldsymbol{y}_g = (y_{g1}, y_{g2}, ..., y_{gD})$. In the process of each memetic evolution, the position of the frog with the worst fitness is adjusted as follows:

$$J_d = r \times (y_{bd} - y_{wd}), \tag{17}$$

$$y_{wd}^{'} = y_{wd} + J_d, J_{max} \geq J_d \geq -J_{max}, \tag{18}$$

where r is a random number between 0 and 1, J_{max} is the maximum allowed change in a frog's position, and $1 \leq d \leq D$. If a better position is produced, the worst frog is replaced by the new solution. Otherwise, repeat Equations (17) and (18) with $\boldsymbol{y}_g$ replacing $\boldsymbol{y}_b$. If there is still no improvement in this case, a solution is randomly generated to replace the worst frog. After a certain number of memetic evolutionary steps of all memeplexes, they are forced to mix and new memeplexes are formed through a shuffling process, and new memetic evolution begins. The algorithm is terminated if the convergence criteria are satisfied.

6.2. Proposed Binary SFLA

SFLA presented above is restricted in real number space. However, many optimization problems are set in a discrete number space. Besides, researchers frequently cast floating-point problem in binary terms and solve them in such a discrete number space. Too meet the need, we propose a binary SFLA for discrete combinatorial problems. Although Eusuff et al. [31] and Rahimi-Vahed et al. [32] have presented SFLA for discrete optimization, their approaches are application-specific while ours is suitable for solving any optimization problem that can be solved by QGA, binary PSO or GAs using binary coding.

In the proposed binary SFLA, every dimension of the frog is constrained to binary values, 0 and 1. In the process of each memetic evolution, the position of the frog with the worst fitness is adjusted as follows:

$$J_d = C \times \left(2\left|y_{bd} - y_{wd}\right| - 1\right), \tag{19}$$

$$\text{sig}(J_d) = \frac{1}{1+\exp(-J_d)}, \tag{20}$$

$$y'_{wd} = \begin{cases} \overline{y}_{wd}, & \text{if } r<\text{sig}(J_d), \\ y_{wd}, & \text{otherwise}, \end{cases} \tag{21}$$

where C is a positive constant, r is randomly generated between 0 and 1, and $\overline{\cdot}$ denotes 'NOT' operator, i.e., $\overline{1}=0$ and $\overline{0}=1$. In this case, if $y_{bd} = y_{wd}$, then $J_d = -C$ and $\text{sig}(J_d)$ will be a small value near 0, thus y'_{wd} may equal to y_{wd} with a high probability. When y_{bd} does not equal to y_{wd}, then $J_d = C$, $\text{sig}(J_d)$ would be a high value near 1 and y'_{wd} may take the value $\overline{y}_{wd}$ (i.e., y_{bd}) with a high probability. So $\text{sig}(J_d)$ denotes the probability of bit y_{wd} being flipped. Note that this memetic evolution will make the frog with the worst fitness approach the one with the best fitness. Higher C means $\boldsymbol{y}'_w = (y'_{w1}, y'_{w2}, \ldots, y'_{wD})$ equal to $\boldsymbol{y}_b = (y_{b1}, y_{b2}, \ldots, y_{bD})$ with a higher probability, while smaller C allows a higher mutation rate. Other than the memetic evolution, the procedure of the binary SFLA proceeds much the same as SFLA presented in Section 6.1.

6.3. Binary SFLA-Based Cognitive Radio Adaptation

As in PSO, the first step to use binary SFLA for cognitive radio adaptation is to map the radio parameters which need to be adjusted to frogs in binary SFLA. The structure of frog $\boldsymbol{y}_i = (y_{i1}, y_{i2}, \ldots, y_{iD})$ is the same as the structure of the position of particles in PSO as shown in Figure 3. As for fitness evaluation, we directly use the objective value computed from Equation (3) as the fitness value of the corresponding frog (solution). The adaptation procedure is as follows:

Step 1: Randomly generate a population of P frogs. Compute the fitness value for each frog. Record the best frog's position $\boldsymbol{y}_g = (y_{g1}, y_{g2}, \ldots, y_{gD})$ in the entire population and its corresponding fitness.
Step 2: Sort the P frogs in order of decreasing fitness value. Partition them into k memeplexes. Each memeplex contains ℓ frogs. Set $\kappa = 1$.
Step 3: For memeplex κ, record the frog with the best fitness $\boldsymbol{y}_b = (y_{b1}, y_{b2}, \ldots, y_{bD})$ and the one with the worst fitness $\boldsymbol{y}_w = (y_{w1}, y_{w2}, \ldots, y_{wD})$. Set $t = 1$.
Step 4: Obtain $\boldsymbol{y}_w^{'}$ according to Equations (19), (20), and (21). If $\boldsymbol{y}_w^{'}$ is better than $\boldsymbol{y}_w$, replace $\boldsymbol{y}_w$ with $\boldsymbol{y}_w^{'}$, otherwise, get a new $\boldsymbol{y}_w^{'}$ according to Equations (19), (20), and (21) but with y_{gd} replacing y_{bd}. If the new $\boldsymbol{y}_w^{'}$ is better than $\boldsymbol{y}_w$, replace $\boldsymbol{y}_w$ with $\boldsymbol{y}_w^{'}$, otherwise, randomly generate $\boldsymbol{y}_w^{'}$ to replace $\boldsymbol{y}_w$. If $\boldsymbol{y}_w^{'}$ is better than $\boldsymbol{y}_b$, replace $\boldsymbol{y}_b$ with $\boldsymbol{y}_w^{'}$, and if $\boldsymbol{y}_w^{'}$ is better than $\boldsymbol{y}_g$, replace $\boldsymbol{y}_g$ with $\boldsymbol{y}_w^{'}$.
Step 5: If t equals to the predefined iterations, go to step 6. Otherwise, set $t = t + 1$, go to step 4.
Step 6: If κ equals to k, go to step 7. Otherwise, Set $\kappa = \kappa + 1$ and go to step 3.

Table 4. Objective values: GA versus QGA versus PSO.

Mode	Algorithm	Generation or Iteration				Standard deviation
		50	100	500	1500	
Mode 1	GA	0.6725	0.7127	0.8405	0.8977	0.0093
	QGA	0.8276	0.8620	0.9116	0.9187	0.0085
	PSO	0.8141	0.8615	0.9091	0.9156	0.0083
Mode 2	GA	0.6686	0.6903	0.7764	0.8183	0.0073
	QGA	0.7767	0.8077	0.8334	0.8360	0.0025
	PSO	0.7606	0.7960	0.828	0.8358	0.0016
Mode 3	GA	0.6894	0.7321	0.8664	0.8912	1.7820×10^{-4}
	QGA	0.8830	0.8890	0.8912	0.8912	1.4207×10^{-4}
	PSO	0.8845	0.8882	0.8911	0.8913	6.0029×10^{-5}
Mode 4	GA	0.5843	0.6060	0.6958	0.7479	0.0015
	QGA	0.6950	0.7248	0.7498	0.7503	1.1703×10^{-5}
	PSO	0.6890	0.7230	0.7485	0.7502	3.6620×10^{-5}

Step 7: If the convergence criteria are satisfied, terminate the algorithm. Otherwise, shuffling all memeplexes, and go to step 2.

When the binary SFLA is terminated, the stored global best frog $\boldsymbol{y}_g = (y_{g1}, y_{g2}, \ldots, y_{gD})$ is the best solution obtained which will be used to configure the radio.

6.4. Population Initialization in Binary SFLA

The frogs in the initial population are generated randomly in the above binary SFLA. However, using information about the problem domain, we can use initial seeding techniques to improve the performance of the binary SFLA. In a quasi-static wireless channel environment, we can assume the environment parameters are changing slowly [13]. In this case, if the weight vector keeps unchanged, the result from the previous cycle would be a good solution in the current situation as well. Denote the solution (frog) obtained at the previous cognition cycle as $\boldsymbol{y}^*$, we can seed a percentage of the initial population with $\boldsymbol{y}^*$ at the current adaptation cycle. The percentage of initial seeding is denoted as $\gamma = P^* / P$, where P^* denotes the number of frogs which are equal to the solution from the previous cognition cycle. Doing this will bias the initial population toward the last decision as shown in Figure 5. The square is the new target optimal solution, the rhombus is the solution obtained at the previous cycle, and the rounds are the randomly generated frogs. The rhombus approximates the target optimum while other initial solutions are spread out randomly to cover the whole search space. As the initial population contains the rhombus, the probability to reach the square is obviously higher than the initial population which only contains the rounds but not the rhombus.

6.5. Simulation Results

In the simulations, we compare our binary SFLA-based adaptation method against the method based on PSO. The population size of the binary SFLA is set 50, and the population is divided into 10 memeplexes, each with 5 frogs. The maximum number of iterations for each memetic evolution in any memeplex is set 5, and the algorithm will be terminated after 30,000 times of fitness evaluations.

For simplicity, mode 1 and mode 2 in Table 2 are used to test the effectiveness of the proposed method. The same simulation environment is used as Section 5.3 except that the constant path loss from the transmitter to the receiver is assumed to be 90 dB here.

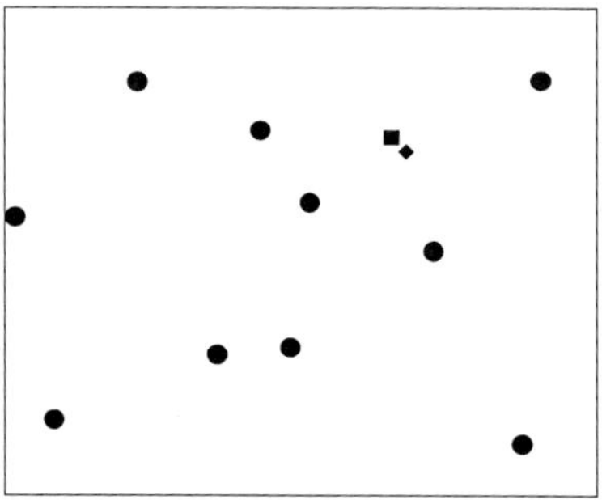

Figure 5. Initial population seeding of binary SFLA.

Table 5. Average objective values: PSO versus binary SFLA.

Fitness evaluation times	Mode 1		Mode 2	
	PSO	Binary SFLA	PSO	Binary SFLA
1,500	0.7810	0.8248	0.5133	0.5434
3,000	0.8302	0.8596	0.5428	0.5585
15,000	0.8699	0.8736	0.5628	0.5636
30,000	0.8731	0.8736	0.5648	0.5648

Table 6. Standard deviation: PSO versus binary SFLA.

Fitness evaluation times	Mode 1		Mode 2	
	PSO	Binary SFLA	PSO	Binary SFLA
1,500	0.0065	0.0081	0.0122	0.0063
3,000	0.0057	0.0029	0.0041	0.0041
15,000	0.0018	1.1703×10^{-16}	0.0018	0.0018
30,000	7.7528×10^{-4}	1.1703×10^{-16}	0.0014	0.0012

Ten independent experiments were conducted to each mode using binary SFLA and PSO respectively. The objective function values of the global best frog in binary SFLA and the global best particle in PSO are recorded and the resulting average objective function values over 10 simulation runs are shown in Figure 6.

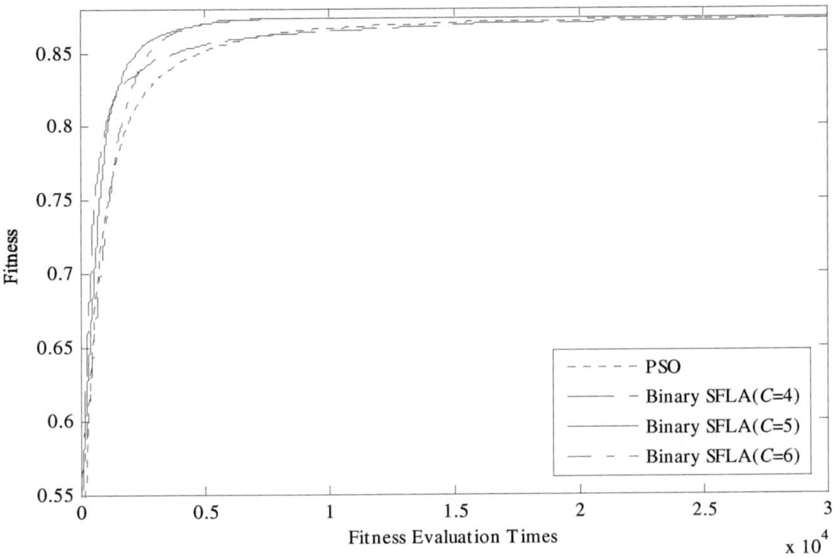

(a)

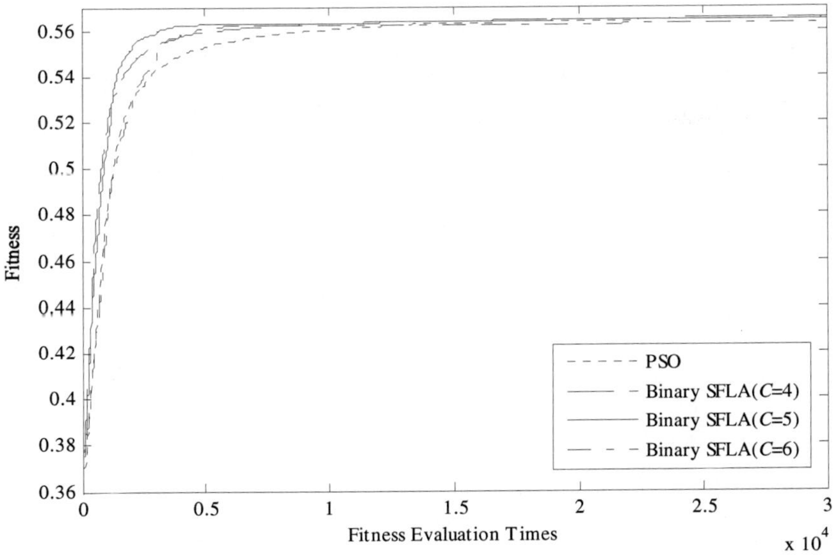

(b)

Figure 6. Performance comparison: PSO versus binary SFLA. (a) mode 1; (b) mode 2.

It illustrates that the binary SFLA performs best in terms of both convergence speed and converged value when $C = 5$ as compared to the other two scenarios where $C = 4$ and $C = 6$, respectively. The exact average objective function values and standard deviations after 1,500, 3,000, 15,000 and 30,000 times of fitness evaluation are shown in Table 5 and Table 6, respectively. It can be obviously seen that the binary SFLA converges faster than the PSO, and the solutions optimized by the binary SFLA have higher objective values (fitness values) than those optimized by the PSO. The standard deviations of the objective values illustrates that the binary SFLA performs as stably as the PSO algorithm.

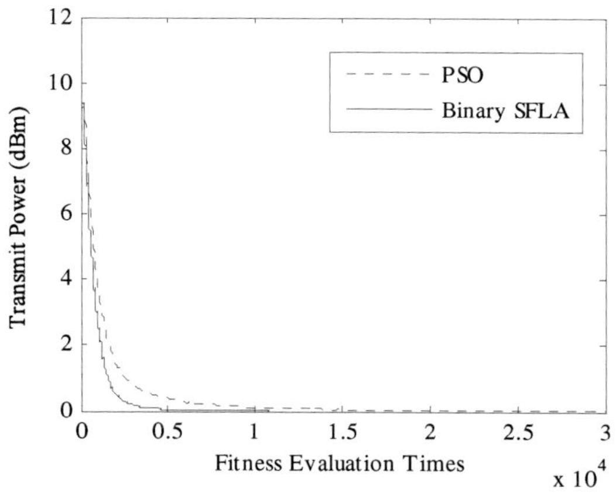

Figure 7. Transmit power convergence: PSO versus binary SFLA. *C*=5.

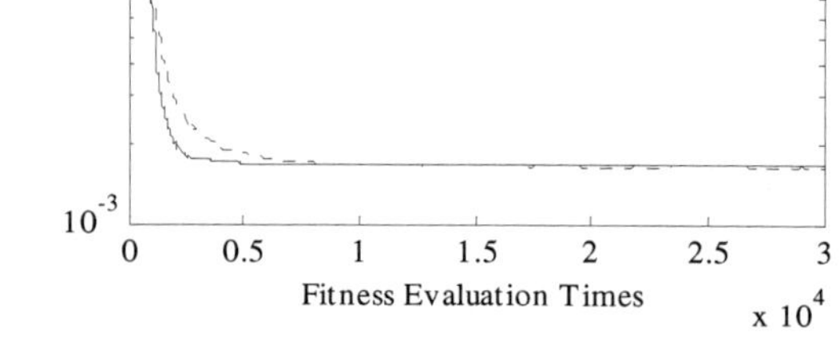

Figure 8. BER convergence: PSO versus binary SFLA. *C*=5.

As a validation of these fitness values, Figure 7 shows the simultaneous average transmit power convergence in mode 1. This plot validates that higher fitness values provide lower transmit powers, and the main objective of mode 1 is optimized. Figure 8 shows the simultaneous average BER over 32 subcarriers in mode 2. It verifies that higher fitness values result in lower BERs. It can also been seen that the main objective of mode 2, minimize BER, is optimized.

Figure 9 illustrates the performance of the initial population seeding method presented in Section 6.4. The weighting scenario is assumed to keep the same between the previous adaptation cycle and the current cycle. We use the concept of environmental variation factor (EVF) which is proposed by Newman et al. [13] to characterize the change in environment parameters but with a simpler definition. We simply define EVF as the ratio of the number of changed parameters to the total number of environmental parameters. Three out of 30 channel attenuation values have changed in this simulation which implies EVF=10%. For the sake of computation complexity, we assume that the solution obtained at the previous cycle corresponds to the global best frog of the binary SFLA after 1,500 times of fitness evaluation. It is shown that the initial seeding technique ($\gamma = 0.1$ or $\gamma = 0.5$) allows the algorithm to start at a higher initial fitness and converge far faster than the standard binary SFLA ($\gamma = 0$). Note that the performance at the early stage of the search is very important for real-time implementation of a cognitive engine, which requires low computation complexity and real-time adaptation. It can also be seen that the binary SFLA with 10% initial seeding with the solution from the previous cycle performs approximately the same as the binary SFLA with 50% initial seeding.

Figure 10 shows the performance of the initial seeding technique where EVF=90%. Similar results can be drawn from Figure 10 as Figure 9. Even though the environmental parameters have changed a lot, as the optimization direction keeps the same, the solution obtained at the previous cycle is still quite a good solution in the current situation. However, comparing Figure 9 (b) and Figure 10 (b), we can see that the initial fitness ($\gamma = 0.1$) in Figure 10 (b) is much smaller than that in Figure 9 (b) due to large EVF.

Figure 9 and Figure 10 show the results when the weight vector keeps unchanged. However, in real situations, the preference information may change over time and the weight vector under the current cycle may be different from the previous cycle. In this case, the solution from the previous cycle might be a bad solution in the current situation. Figure 11 illustrates the performance of the initial seeding technique when the weight scenario changes from mode 1 to mode 2 and vice versa. The main objective of mode 1 and mode 2 are minimizing transmit power and minimizing BER, respectively. As the solution which minimizes transmit power has a relative high BER, these two objectives contradict each other, thus a good solution in mode 1 will be a bad solution in mode 2 and vice versa. It is shown in Figure 11 that with 10% or 50% initial seeding, the algorithm performs as well as the standard binary SFLA ($\gamma = 0$). However, there is no advantage of the initial seeding over the standard binary SFLA when the weighting scenario changes. An alternative to keep the advantage of the initial seeding technique is to save the latest solution in each weighting mode.

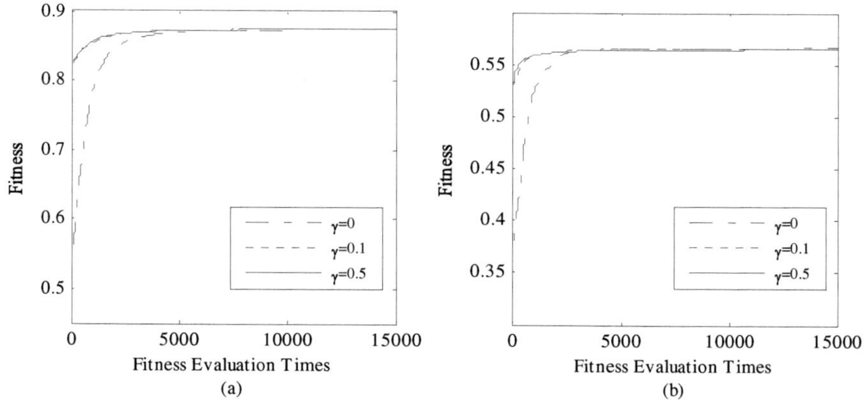

Figure 9. Performance of the initial seeding technique. EVF=10%. *C*=5. (a) mode 1; (b) mode 2.

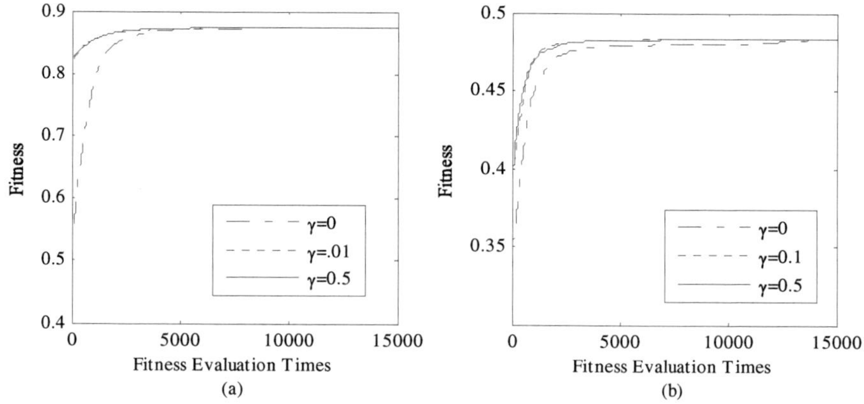

Figure 10. Performance of the initial seeding technique. EVF=90%. *C*=5. (a) mode 1; (b) mode 2.

When cognitive engine encounters a specific weighting scenario, it uses the latest solution under this weighting scenario to seed the current initial population. This can be beneficial and feasible when the number of possible weighting scenarios is not too large.

7. Challenges and Future Notes

Evolutionary algorithms serve as a potential solution for cognitive radio adaptation and optimization problem. There are still some challenges which need to be tackled for fully implementation of evolutionary algorithms in cognitive radios, including cross-layer adaptation, dynamic environment consideration and black-box modeling.

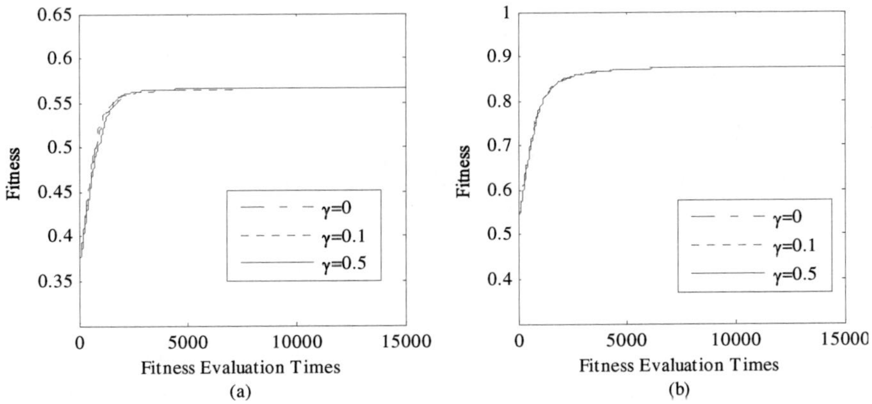

Figure 11. Performance of the initial seeding technique. EVF=10%. *C*=5. (a) The weighting scenario changes from mode 1 to mode 2; (b) The weighting scenario changes from mode 2 to mode 1. In both figures, three curves approximately overlap one another.

7.1. Cross-Layer Adaptation and Optimization

Adaptation itself is a complex problem in cognition cycle, because cognitive radio needs to consider several parameters and requirements simultaneously. The requirements and the parameters fall into different layers such as transmit power on the PHY layer, channel coding rate and data rate on the data link layer, and clustering parameters on the network layer. Therefore, cognitive radio needs an overall adaptation that covers multiple layers [33,34]. The simulations presented in this chapter focus much on the PHY layer. Cross-layer design is an emerging area, and applying evolutionary algorithms for solving fully cross-layer adaptation problem still needs further investigation.

7.2. Dynamic Environment Consideration

Cognitive radio adaptation is dynamic in nature due to the changing RF environment, users' needs, and hardware conditions. The optimal solution will change as the objective functions, environment parameters or constraints change. One simple method to deal with such dynamic environment is to restart the search process each time the environment changes and search the new solution from the start. However, this is impractical not only because it is a waste of time that not using the past information but also it will consume a lot of computation resources when the environment changes fast. Although the initial population seeding method presented in Section 6.4 uses the past information, its application is still limited in that it is only effective when the optimization direction (weight vector) keeps the same. Therefore, new evolutionary algorithms are needed. Dynamic evolutionary algorithms such as primal-dual genetic algorithm [35] may be a good choice in this situation.

7.3. Black-Box Modeling

In many CR proposals including Section 5.3.1 in this chapter, analytical models have been used for performance characterization and deriving some objective functions. However, in several situations an analytical model might not be available and/or practical to use. In such situations, black-box modeling may be a solution. For instance, Baldo et al. [36] proposed to use neural networks to model cognitive radio systems and build relationships between the adjustable parameters and objective functions. Evolutionary algorithms suitable for black-box modeling situation need to be further investigated.

8. Conclusion

The primary goal of the cognitive engine is to optimize the radio, and the secondary functions are to observe and learn in order to provide the knowledge required to perform the adaptation [2]. This chapter has focused on the adaptation and optimization of the radio. Four evolutionary algorithms, classic GAs, QGA,

PSO and SFLA, have been investigated thoroughly for adaptation. It is shown by simulation that SFLA may be the best choice. However, evolutionary algorithms are evolving themselves, and new methods such as predatory search strategy [37] are encouraged to be investigated for the adaptation problem. Furthermore, evolutionary algorithms for cross-layer adaptation, dynamic environment, and black-box modeling situation still need further investigation.

References

[1] Haykin, S. (2005). Cognitive radio: brain-empowered wireless communications. *IEEE J. Sel. Area. Comm.*, vol. 23, 201-220.

[2] Rondeau, T. W. & Bostian, C. W. (2006). Cognitive techniques: physical and link layers. In *Cognitive Radio Technology*; In B. Fette (Ed.), Elsevier: Burlington, MA, 219-266.

[3] Arslan, H. & Yarkan, S. (2007). Cross-layer adaptation and optimization for cognitive radio. In *Cognitive Radio, Software Defined Radio, and Adaptive Wireless Systems*; In H. Arslan (Ed.), Springer: Dordrecht, AA, 426.

[4] Xing, Y., Kushwaha, H., Subbalakshmi, K. P. & Chandramouli, R. (2007). Codes and games for dynamic spectrum access. In *Cognitive Radio, Software Defined Radio, and Adaptive Wireless Systems*; In H. Arslan (Ed.), Springer: Dordrecht, AA, 161.

[5] Rondeau, T. W., Rieser, C. J. & Bostian, C. W. (2004). Cognitive radios with genetic algorithms: intelligent control of software defined radios. *Software Defined Radio Forum Technical Conference*, C3-C8.

[6] Rieser, C. J. (2004). *Biologically inspired cognitive radio engine model utilizing distributed genetic algorithms for secure and robust wireless communications and networking*. Ph. D. Dissertation, Virginia Polytechnic Institute and State University.

[7] Rondeau, T. W., Le, B., Maldonado, D., Scaperoth, D. & Bostian, C. W. (2006). Cognitive radio formulation and implementation. In *Proceedings of the First International Conference on Cognitive Radio Oriented Wireless Networks and Communications (CROWNCOM)*.

[8] Hauris, J. F. (2007). Genetic algorithm optimization in a cognitive radio for autonomous vehicle communications. In *Proceedings of the 2007 IEEE International Symposium on Computational Intelligence in Robotics and Automation*, 427-431.

[9] Newman, T. R., Barker, B. A., Wyglinski, A. M., Agah, A., Evans, J. B. & Minden, G. J. (2007). Cognitive engine Implementation for wireless

multicarrier transceivers. *Wirel. Commun. Mobile Comput.*, vol. 7, 1129-1142.

[10] Baynast, A. D., Mähönen, P. & Petrova, M. (2008). ARQ-based cross-layer optimization for wireless multicarrier transmission on cognitive radio networks. *Comput. Netw.*, vol. 52, 778-794.

[11] Zhao, Z., Zheng, S. & Xu, C. (2007). Cognitive engine implementation using genetic algorithm and simulated annealing. *WSEAS T. Commun.*, vol. 6, 773-777.

[12] Zhao, Z., Zheng, S., Shang, J. & Kong, X. (2007). A study of cognitive radio decision engine based on quantum genetic algorithm. *Acta Phys. Sin.*, vol. 56, 6760-6766.

[13] Newman, T. R., Rajbanshi, R., Wyglinski, A. M., Evans, J. B. & Minder, G. J. (2007). Population adaptation for genetic algorithm-based cognitive radios. In *Proceedings of the Second International Conference on Cognitive Radio Oriented Wireless Networks and Communications (CROWNCOM).*

[14] Zhao, Z., Xu, S., Zheng, S. & Shang, J. (2009). Cognitive radio adaptation using particle swarm optimization. *Wirel. Commun. Mobile Comput.*, vol. 9, 875-881.

[15] Elbeltagi, E., Hegazy, T. & Grierson, D. (2005). Comparison among five evolutionary-based optimization algorithms. *Adv. Eng. Inform.*, vol. 19, 43-53.

[16] Eusuff, M. M. & Lansey, K. E. J. (2003). Optimization of water distribution network design using the shuffled frog leaping algorithm. *Water Resour. Planning Manage.*, vol. 129, 210-225.

[17] Ishibuchi, H. & Murata, T. (1998). A multi-objective genetic local search algorithm and its application to flowshop scheduling. *IEEE T. Syst. Man CY. C.*, vol. 28, 392-403.

[18] Goldberg, D. E. (1989). *Genetic Algorithms in Search, Optimization, and Machine Learning*; Addison-Wesley Publishing Co.: Reading, MA.

[19] Holland, J. H. (1975). *Adaptation in Natural and Artificial Systems*; Univ. Of Michigan Press: Ann Arbor, Mich.

[20] Srinivas, M. & Patnaik, L. M. (1994). Genetic algorithms: a survey. *Computer.*, vol. 27, 17-26.

[21] Narayana, A. (1997). An introductory tutorial to quantum computing. In *Proceedings of IEE Colloquium on Quantum Computing: Theory, Applications and Implications*, 1/1-1/3.

[22] Hey, T. (1999). Quantum computing: an introduction. *Comput. Control Eng., vol. 10*, 105-112.

[23] Han, K. H. & Kim, J. H. (2000). Genetic quantum algorithm and its application to combinatorial optimization problems. In *Proc. of the 2000 IEEE conference on Evolutionary Computation*, 1354-1360.
[24] Yang, J. & Zhuang, Z. (2003). Multi-universe parallel quantum genetic algorithm and its application to blind source separation. In *Proc. of IEEE International Conference on Neural Network and Signal Processing*, 14-17.
[25] Kennedy, J. & Eberhart, R. (1995). Particle swarm optimization. In *Proc. of IEEE International Conference on Neural Networks*, 1942-1948.
[26] Shi, Y. & Eberhart, R. (1998). A modified particle swarm optimizer. In *Proc. of the IEEE International Conference on Evolutionary Computation*, 69-73.
[27] Kennedy, J. & Eberhart, R. (1997). A discrete binary version of the particle swarm algorithm. In *Proc. of the Conference on Systems, man, and Cybernetics*, 4104-4109.
[28] Kennedy, J. (1997). The particle swarm: social adaptation of knowledge. In *Proceedings of the IEEE International Conference on Evolutionary Computation*, 303-308.
[29] Liao, C., Tseng, C. & Luarn, P. (2007). A discrete version of particle swarm optimization for flowshop scheduling problems. *Comput. Oper. Res.*, vol. 34, 3099-3111.
[30] Proakis, J. G. (2000). *Digital Communications, fourth ed.,* McGraw-Hill: New York, NY.
[31] Eusuff, M., Lansey, K. & Pasha, F. (2006). Shuffled frog-leaping algorithm: a memetic meta-heuristic for discrete optimization. *Eng. Optim.*, vol. 38, 129-154.
[32] Rahimi-Vahed, A. & Hossein Mirzaei, A. (2008). Solving a bi-criteria permutation flow-shop problem using shuffled frog-leaping algorithm. *Soft. Comput.*, vol. 12, 435-452.
[33] Arslan, H. & Yarkan, S. (2007). Cross-layer adaptation and optimization for cognitive radio. In *Cognitive Radio, Software Defined Radio, and Adaptive Wireless Systems*; In H. Arslan (Ed.), Springer: Dordrecht, AA, 421-456.
[34] Srivastava, V. & Motani, M. Cross-layer design and optimization in wireless networks. In *Cognitive Networks*; In Q. H. Mahmoud (Ed.), Wiley: Chichester, WS, 121-146.
[35] Yang, S. (2003). Non-stationary problem optimization using the primal-dual genetic algorithm. In *Proc. of 2003 Congress on Evolutionary Computation*, vol. 4, 2246-2253.

[36] Baldo, N. & Zorzi, M. (2008). Learning and adaptation in cognitive radios using neural networks. In *Proc. of IEEE CCNC 2008*, 998-1003.
[37] Linhares, A. (1999). Synthesizing a predatory search strategy for VLSI layouts. *IEEE T. Evolut. Comput.*, vol. 3, 147-152.

Reviewed by Xiaoniu Yang, JiangNan Electronic Communication Research Institute, Jiaxing 314033, China.

In: Cognitive Radio
Editors: P.D. McGuire et al, pp. 73-99
ISBN: 978-1-60876-604-8

Chapter 3

MODULATION TECHNIQUES FOR COGNITIVE RADIO IN OVERLAY SPECTRUM SHARING ENVIRONMENT

I. Budiarjo[a], H. Nikookar[b] and L.P. Ligthart[c]
International Research Centre for Telecommunications and Radar (IRCTR),
Department of Electrical Engineering, Mathematics,
and Computer Science, Delft University of Technology, Mekelweg 4,
2628 CD Delft, The Netherlands.

Abstract

The growing demand on wireless communication systems to provide high data rates has brought with it the need for a flexible and efficient use of the spectrum resource, which is a scarce commodity. The regional spectrum allocation policy counteracts the free mobility of radio communication equipment. The vast majority of the available spectral resources have already been licensed, so it appears that there is little or no room to add any new services, unless some of the existing licenses are discontinued. Furthermore, recent studies and measurements have shown that vast portions of the licensed spectra are rarely used due to the inflexible spectrum regulations. The whole idea behind Cognitive Radio (CR) use is that it should prompt effective

[a] E-mail address: i.budiarjo@irctr.tudelft.nl, Tel: +31 (0) 15 27 87860, Fax: +31 (0) 15 27 84046. (Corresponding author).
[b] E-mail address: h.nikookar@irctr.tudelft.nl.
[c] E-mail address: l.p.ligthart@irctr.tudelft.nl.

spectrum use, since intelligence and learning processes aid the radio system to access the spectrum effectively. The CR system has learning and understanding capabilities so that the stated goals may be achieved. In this paper we shall limit the scope of cognition to reduce mutual interference between CR-based rental (unlicensed) users and licensed users (LUs) and in providing coexistence between them. It is expected that the rental users will be allowed to transmit and receive data over portions of spectra when primary (i.e., licensed) users are inactive. In this article we will introduce the modulation strategies employed to realize a co-existence between the CR-based rental system and the licensed system. This is done in such a way that the rental users (RUs) are invisible to the licensed users (LUs). We consider the rental user accesses the unoccupied LU band in overlay fashion.

1. Introduction

Cognitive Radio (CR) as defined in [1] is an intelligent wireless communication system that is aware of its surrounding environment and uses that understanding by building to learn from the environment and adapt its internal states to variations in incoming radio frequency (RF) stimuli by changing its operating parameters in such a way that highly reliable communication and efficient utilization of the radio system can be achieved. The key words in this definition are awareness, intelligence, learning, adaptivity, reliability, and efficiency. Mitola in [2] developed the notion that CR signifies a radio that employs model-based reasoning to achieve a specified level of competence in radio related domains. This includes supervised and unsupervised learning, and it is assumed to be implemented by software radio or software defined radio architecture.

In [2] spectrum pooling is addressed as a mechanism where the current spectrum owners allow portions of their spectrum to be utilized by rental users (RUs) which apply the CR system in their transmission for certain periods. This mechanism is applied to consider the actual spectrum occupancy condition where the allocated spectrum is not occupied most of the time. In fact some portions of the spectrum are rarely used [3]. The spectrum owners, who are known as the licensed users (LUs) are ordinary mobile terminals and their associated base stations. They do not possess much intelligence. The RUs, on the other hand, should possess the intelligence to sense the spectrum and use whatever resources are available when they need them. At the same time, the RUs should renounce the spectrum when an LU begins transmission. The RU access should be done in such a way that the neighboring LUs will not be aware of its existence. This can be accomplished by assuring that the neighboring LUs will not be interrupted or interfered with as the RU accesses the spectrum hole adjacent to the LU's band. This stipulation can be technically interpreted as in overlay opportunistic RU (CR) access mode where part of the RU spectrum which is located on the neighboring occupied LU band should be shaped to be low as possible so that it will not significantly interrupt the LU transmission as well as reducing mutual interference between the RU and LU, hence reliable communications can be accomplished. This shaping can be interpreted as a sidelobe suppression technique. In this article the techniques used to fulfill the co-existence between LU and RU in opportunistic spectrum access and to also achieve the CR objectives are described. We shall review the multicarrier and single carrier approach using either the Fourier or wavelet basis function as the CR modulation technique. The remainder of the manuscript is organized as follows. The multi carrier cognitive radio system is described in section 2. Section 3 is devoted to single carrier cognitive radio system, and the summary is provided in section 4.

2. Multicarrier Cognitive Radio System

Orthogonal frequency division multiplexing (OFDM) is well known as a spectrally efficient modulation mode due to its overlapped carrier spectrum which still remain orthogonal to each other, and it gives a higher bit rate compared to the single carrier transmission. Recently research interest has been developed in replacing the static sines/cosines basis function in OFDM with wavelet bases due to its flexibility and adaptability which can be tailored to satisfy the engineering demand. The next subsections will describe how these two multicarrier modulation techniques can be applied in cognitive radio systems.

2.1. OFDM Based Cognitive Radio System

OFDM has recently been introduced as a strong multicarrier modulation scheme candidate to be applied in CR [4][5]. The rationale is that any CR system needs to sense the spectrum, and that involves some sort of spectral analysis. FFT can be used for spectral analysis while at the same time acting as an OFDM demodulator. It has therefore been suggested that OFDM should be seen as the proper candidate for multicarrier-based CR systems. OFDM also has the capability to notch the parts of its carriers which are coincidentally within the region of the LU's band. Such OFDM flexibility simplifies the application of CR dynamic spectrum access. The OFDM carriers notching technique is introduced in [4][5] and it is called "Spectrum Pooling". The shortcomings of the OFDM root back to the large side-lobes of the frequency response of filters that characterize the channel associated with each subcarrier. The large side lobes result in significant interference between the subcarriers originating from different RUs and between LUs and RUs.

The sidelobes of the Fourier based OFDM can be observed from its power spectral density (PSD),

$$PSD(f) = \frac{1}{N_{FFT}} \left| \sum_{m=0}^{N_{FFT}-1} \sqrt{p_m} X_m \int_{-(1+\alpha)\frac{T_u}{2}}^{(1+\alpha)\frac{T_u}{2}} g(t) e^{-j2\pi(f-f_m)t} dt \right|^2 \quad (1)$$

where N_{FFT} is the number of FFT points, T_u is the useful signal duration, $g(t)$ is the window function, and α is the roll off factor of the window, while f_m, p_m and X_m are the respective frequency, allocated power and the symbol from QAM or PSK mapping on the subcarrier m. Equation (1) shows that parameters such as allocated power (p_m), symbol amplitude (S_m) and window ($g(t)$) can be set to resolve the sidelobes problem.

2.1.1. Sidelobe Suppression by Windowing and Carriers Deactivation

In [4] the extension of each OFDM block with long cyclic prefix and suffix samples and the application of some windowing to reduce the side-lobes of the subcarrier channels is proposed. Obviously, this solution occurs at the expense of bandwidth loss because excessive time has to be allocated to cyclic extensions which otherwise could be used for data transmission. Furthermore in [4], the interference reduction method is extended by deactivating subcarriers located adjacent to the LU's band which provides a kind of shield to the LUs. One commonly used window is the raised cosine window [4]. The other available forms are Bartlett [6], Better than Raised Cosine (BTRC) [7], and Flipped Inverse Hyperbolic Secant (farchsech) [8] windows. Of these windows, it is the farchsech window that provides the lowest sidelobes. The window is designed according to (2),

$$g(t)=\begin{cases}\dfrac{1}{T_u}, & 0\le|t|\le\dfrac{T_u(1-\alpha)}{2}\\ \dfrac{1}{T_u}\left[1-\dfrac{1}{\alpha T_u\gamma}\operatorname{arcsech}\left(\dfrac{1}{\alpha T_u}\left(\dfrac{T_u(1+\alpha)}{2}-|t|\right)\right)\right], & \dfrac{T_u(1-\alpha)}{2}\le|t|\le\dfrac{T_u}{2}\\ \dfrac{1}{T_u}\left[\dfrac{1}{\alpha T_u\gamma}\operatorname{arcsech}\left(\dfrac{1}{\alpha T_u}\left(|t|-\dfrac{T_u(1-\alpha)}{2}\right)\right)\right], & \dfrac{T_u}{2}\le|t|\le\dfrac{T_u(1+\alpha)}{2}\\ 0 & \text{Otherwise}\end{cases} \quad (2)$$

where $\gamma=2\ln(\sqrt{3}+2)/(\alpha T_u)$, α is the roll-off factor, and T_u is the useful OFDM symbol interval (without guard interval). The shape of the window in the region of $\frac{T_u(1-\alpha)}{2}\le|t|\le\frac{T_u(1+\alpha)}{2}$ determines how rapidly the OFDM spectrum falls to zero. The PSD will hit zero on the frequencies in the interval of $1/T_u$.

According to [9] the duration of the OFDM signal should be $2T_u$ in order to complete a total $2N_{FFT}$ samples to preserve the orthogonality, and zeros are added in the region outside the window $g(t)$. If the duration of the OFDM signal is only T_u $(1+\alpha)$ then the window will affect the transmitted signal, and as a consequence an error floor will be formed. Figure 1(a) shows an example of a Bartlett window with an OFDM signal $T_u(1+\alpha)$ duration which affects the transmitted signal. In an effort to avoid the error floor, the applied window must not influence the signal during its effective period [10]. To fulfill this requirement the window forms are expanded, so that new T_u becomes $2T_u$, and α is restricted in the range of $0\le\alpha\le(1-T_{GI}/T_u)/2$, where T_{GI} is the guard interval duration. The orthogonality is preserved since the PSD hits zeros in the $1/(2T_u)$ interval at the expense of the longer duration of the OFDM signal $(2T_u(\alpha+1))$. The solution is that the window is truncated to fit the $2T_u$ OFDM duration, as depicted in Figure 1(b) [11]. The orthogonality is preserved by applying the rectangular receiver filter with T_u duration, that is implemented by DFT [10]. In this way there is more freedom to shape the window in the $T_u(1-\alpha)\le|t|\le T_u$ region as long as its PSD on the symbol boundaries remains low, e.g Gaussian window or half sine window [12].

By referring to (1) the deactivating carriers adjacent to the licensed band will provide flexible guard bands that will make the PSD sidelobes of the RU's OFDM signal on the licensed band lower. In OFDM bits errors are typically concentrated in a set of severely faded subcarriers. This is due to the channel propagation condition. For this reason, it is possible that the quality of service may not be maintained. Meanwhile, the carriers resource is limited due to the deactivation of the RU carriers. Furthermore, the long OFDM symbol duration due to the windowing reduces the OFDM transmission bit rate. Having adaptive OFDM in CR is thus a solution. The heavily faded and noisy subcarriers are excluded from transmission to improve the overall bit error rate (BER) of the system, and the loss of throughput is counteracted by applying higher order modulation modes to the subcarriers which have better signal to noise ratios (SNR) or BER. In [11] the combination of adaptive bit loading and spectrum pooling has been evaluated. The window was designed in accordance with Figure 1(b). The results show how adaptive bit loading algorithms can outperform the ordinary OFDM.

Recently in [13] the overlapped OFDM symbol transmission with long symbol duration (known as offset OFDM) has been introduced to counteract the throughput loss. The scheme is described in Figure 2. The delay between one OFDM symbol and another is designed in such a way that no intersymbol interference occurs, which means that the next OFDM symbol should start after the end of the useful data part of the previous OFDM symbol. If zeros are inserted at the prefix and suffix points (outside the T_{GI} and T_U area as depicted in Figure 1(b)), the useful data of the next OFDM symbol should start after the last useful data of the previous OFDM symbol.

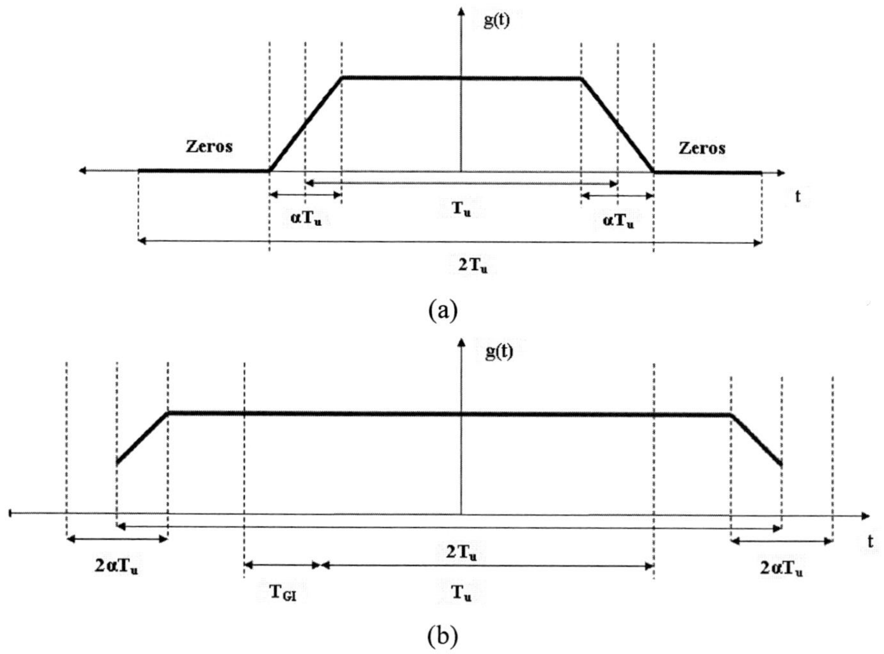

Figure 1. (a) A window design that influences the transmitted signal (b) A window design that does not influence the transmitted signal [11].

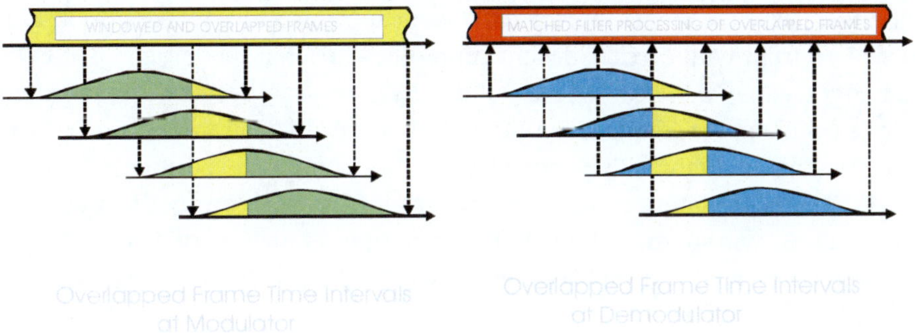

Figure 2. Overlapped OFDM symbol with zero ISI. The figure on the left reflects the technique applied at the transmitter point while the figure on the right is applied at the receiver point.

2.1.2. Sidelobe Suppression by Cancellation Carriers Insertion

In [14]-[16] instead of having deactivated carriers adjacent to the LU band, these carriers are used to cancel out the sidelobes on the LU band. The concept is to find the proper amplitude of the cancellation carriers in such a way that the resulting sidelobes be as low as possible. The optimization can be formulated as:

$$|X(f)|^2 = \frac{1}{N_{FFT}} \left| \sum_{\substack{m=1 \\ m \neq cc}}^{N_{FFT}} \sqrt{p_m}(X_m) \int_{-(1+\alpha)T_u}^{(1+\alpha)T_u} g(t) e^{-j2\pi(f-f_m)t} dt + \sum_{\substack{cc=cc1 \\ cc \neq m}}^{ccN} \sqrt{p_{cc}}(Xc_{cc}) \int_{-(1+\alpha)T_u}^{(1+\alpha)T_u} g(t) e^{-j2\pi(f-f_{cc})t} dt \right|^2 \quad (3)$$

$$\textit{subject to} \quad |X(f_{LU})|^2 < P_{Threshold}, \quad \textit{and} \quad \sum_{m=1}^{N_{FFT}} p_m = P_T$$

where ccN is the total number of sidelobes cancellation carriers, and Xc_{cc} is the symbol on the cancellation carrier position with the range within $cc1,\ldots,ccN$. The power constraint for all subcarriers including the cancellation carriers remains thus, no extra energy is added. The possible high power required by the cancellation carriers can be compensated from the unused power due to the carriers deactivation.

We introduce an alternative optimization problem by simplifying it to set of ccN linear equations to solve ccN cancellation carrier amplitude values. Figure 3 provides a simple example of how the cancellation carriers with this new optimization problem work. The spectrum is derived from 9 subcarriers (including the cancellation carriers) and uses a rectangular window which produces a spectrum from the sum of the sinc functions. One cancellation carrier is located at the left edge of the RU band adjacent to the LU band and another one is at the right edge. They are signed by the dark vertical arrows. If we take a look at the OFDM spectrum with carriers deactivation (one carrier at the left and one carrier at the right edge) denoted by the solid curve, we can see that the maximum sidelobe occurs at the frequency point half of carrier spacing from the deactivated carrier which, in Figure 3 is indicated by the coordinate position (X:-4.5,Y:-0.1564). The cancellation carriers can be weighted in such a way that the sum of their spectras will nullify the OFDM spectrum at that particular coordinate position (e.g. by having a cancellation carrier spectrum coordinate ($X_{c1}=X_{c2}=-4.5$, $Y_{c1}=0.1477$, $Y_{c2}=0.0087$)).

The same technique is also applied to the sidelobes reduction on the right side. The resulting spectrum derived from using the cancellation carriers technique is depicted by the dashed lines. The sidelobes are lower than the sidelobe of the OFDM with the deactivated carriers.

The cancellation carriers can be combined with windowing [17]-[18]. Different windows provide different sidelobe levels. Combining a window that gives very low sidelobes (e.g. the farchsech window) with the cancellation carriers technique will lead to a very significant sidelobe (interference) reduction in the LU band. The results in [17] have shown that the combination of 2 cancellation carriers on each side adjacent to LU band and raised cosine window with roll off factor (α) 0.2 gives more than 30 dB sidelobes reduction compared to the conventional OFDM.

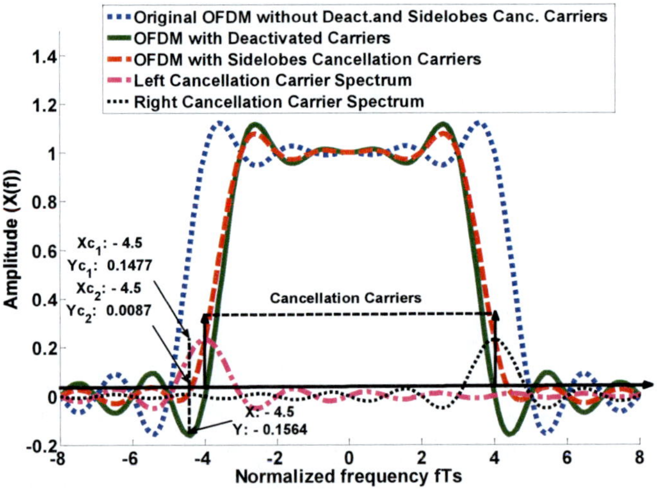

Figure 3. The OFDM spectrum amplitude with the effect of the deactivated carriers or cancellation carriers.

2.1.3. Sidelobe Suppression by Subarrier Weighting

Subcarrier weighting is introduced in [19]. The weighting can be seen as varying the allocation power p_m on each carrier m which, obviously, according to (1) will affect the spectrum amplitude. By selecting the appropriate weight for each of the carriers the desired low sidelobes can be obtained. The sidelobe reduction by means of carriers weighting is attained at the expense of the bit error rate (BER) versus signal to noise ratio (SNR) degradation [19], since the weighting will give unequal amounts of transmission power on each of the OFDM carriers. The weighting method was originally used in [20] to counteract the high peak to average power ratio (PAPR) problem in OFDM.

2.1.4. Sidelobe Suppression by Multiple Choice Sequence

In [21] a multiple choice sequence is introduced, where the multiple choice sequence abiding by the symbol constellation approach, the interleaving approach and the phase approach are discussed. The symbol constellation approach is applied by mapping the symbol S_m in (1) into $S'_m\ (l)$ which is still derived from one of the CS symbols after the QAM or PSK mapping that was previously used to produce S_m . CS is the constellation size of the QAM or PSK mapping. The index l ranges from $l=1,2,...,L$, where L is the number of possible alternative values. This means that the symbol S_m can be mapped into L possible symbols. By putting the N_{FFT} symbols of $S'_m\ (l)$ into (1), and observing from all L possible PSDs, the one with the lowest sidelobe is then chosen for transmission. An example of simple mapping from S_m into $S'_m\ (l)$ is given in [21]. The CS points of QAM or PSK are numbered as 0,1,…,CS-1, hence the symbol S_m is assigned to the number I_m ($I_m \in \{0,1,...,CS-1\}$) according to its point position. The $S'_m\ (l)$ will be the symbol value on the new number $I_m(l)$ which is derived from the addition of the assigned number I_m with the random number $D_m(l)$ which has L different possible values [21],

$$I_m(l) = I_m + D_m(l). \tag{4}$$

By transmitting the information about the set index information l, the receiver can decode the received signal and reconstruct the original S_m symbol. By means of the interleaving approach, the N_{FFT} S_m symbols are interleaved by L different interleaver rules,

$$\mathbf{S'(l)} = \mathbf{\Pi(l)S}, \tag{5}$$

where $\boldsymbol{\Pi}(l)$ is the permutation matrix that will do the interleaving. The matrix $\boldsymbol{\Pi}(l)$ has L different possible matrix values. The permutation matrix $\boldsymbol{\Pi}(l)$ will be available at the transmitter and the receiver. Among L different possible OFDM spectrum using set of symbols $\boldsymbol{S}'(l)$, the one with the lowest sidelobes is chosen for transmission. The receiver will use the deinterleaver permutation matrix $\boldsymbol{\Pi}^{1}(l)$ to recover the transmitted data upon receiving the side information index l.

In the phase approach each phase of symbol S_m is shifted by random phase $\theta_n(l)$ with L different possible values. The phase shifted symbol $S'_m(l)$ becomes [21],

$$S'_m(l) = S_m e^{j\theta_n(l)} \tag{6}$$

where $\theta_n(l)$ is in the range of between 0-2π. In the vector notation, the phase shifted signal $\boldsymbol{S}'(l)$ is derived from point to point multiplication between the original signal $\boldsymbol{S}$ and the complex number vector derived from $e^{j\boldsymbol{\theta}(l)}$. As in the constellation and interleaver approach, in the phase approach, among L possible sets of vector $\boldsymbol{\theta}(l)$, the one that produces the lowest sidelobes is chosen and applied to the symbol vector $\boldsymbol{S}$. The index information l is transmitted so that the receiver can reconstruct the symbol vector $\boldsymbol{S}$. Apart from the random shift, random amplitude shift can also be applied, but in such cases the BER performance will be affected. Another approach is to only apply the multiple choice sequence technique to the carriers adjacent to the LU band, since these carriers affect the sidelobes magnitude most [21].

2.1.5. Sidelobe Suppression by Adaptive Symbol Transition

Sidelobes Suppression by time domain approach has been introduced in [22]. Time domain symbol transition is inserted between two OFDM symbol, and the symbol transition value is obtained in such a way that the signal spectrum or the Fourier transform of the two consecutive time domain OFDM symbol (including the symbol transition) will have low sidelobes. The objective is to minimize the spectrum resulting from two consecutive time domain OFDM symbol without the symbol transition added by the spectrum contributed from the symbol transition on the LU band [22]. If the OFDM based cognitive radio signal contribution on the LU is described as

$$\mathbf{I} = \mathbf{F}_{LU} \begin{bmatrix} \mathbf{y}^{n} \\ \mathbf{a}^{n} \\ \mathbf{y}^{n+1} \end{bmatrix} \tag{7}$$

where $\boldsymbol{I}$ is the signal part of the cognitive radio signal in LU band, $\boldsymbol{F_{LU}}$ is the Fourier transformation matrix for the signal in LU band, y^n is the nth time domain OFDM symbol including the guard interval, $\boldsymbol{a}^n$ is the additive symbol transition, and y^{n+1} is the following OFDM symbol, then the optimization problem will be to find the vector $\boldsymbol{a}^n$ in such a way that $\boldsymbol{I}$ will be as small as possible. The symbol transition scheme is depicted in Figure 4.

The symbol transition will not introduce any inter symbol interference (ISI) but the transmission duration becomes longer, which means that the bit rate is reduced. The search for the time domain symbol transition values are applied by the linear least squares problem with a quadratic inequality constraint [22].

2.1.6. Sidelobe Suppression by Frequency Domain Additive Signal

In [23] sidelobe suppression for OFDM based cognitive radio can be achieved by mapping the data symbol using QPSK modulation mode and add the frequency domain symbol vector $\boldsymbol{X}$ with random values vector $\boldsymbol{d}$ in such a way that the resulting PSD in the LU region be as small as possible. The optimization problem becomes

$$|X(f)|^2 = \frac{1}{N_{FFT}} \left| \sum_{m=1}^{N_{FFT}} \sqrt{p_m}(X_m + d_m) \int_{-(1+\alpha)T_u}^{(1+\alpha)T_u} g(t) e^{-j2\pi(f-f_m)t} dt \right|^2 \tag{8}$$

$$\textit{subject to} \quad |X(f_{LU})|^2 < P_{Threshold}, \quad \textit{and} \quad \sum_{m=1}^{N_{FFT}} p_m = P_T$$

where d_m is the additive signal on subcarrier n. If $X'_m = X_m + d_m$ then in order not to destroy the cognitive radio BER performance and keep the total allocated power constant, the value of d_m is restricted to a circle with magnitude $|\, d_m \,| \leq 0.7$ [23]. In this way, according to the QPSK gray mapping, the value of X'_m will not give decision error at the receiver, except in the severely noise and faded channel. Simulation results in [38] have shown about 15 dB sidelobes reduction could be achieved with the cost of 2.8 dB SNR degradation as the upper limit value of $|\, d_m \,|$ is 0.7.

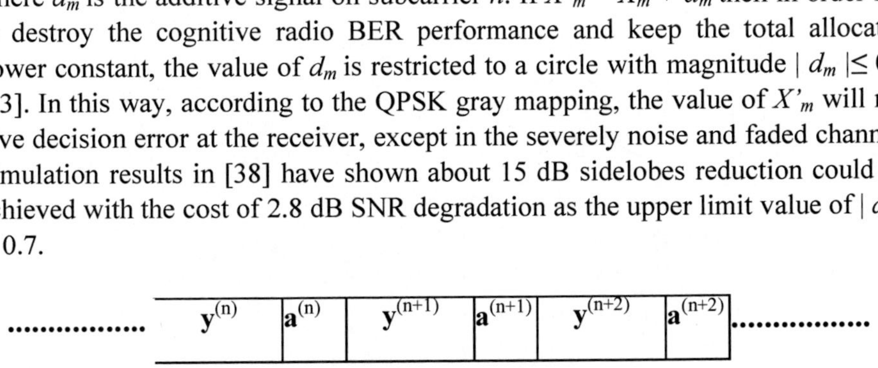

Figure 4. Adaptive symbol transition scheme.

2.1.7. Sidelobe Suppression by Constellation Expansion

The most recent technique in OFDM sidelobe suppression has been proposed in [24], where the M-PSK modulation mapping is expanded to 2xM-PSK mapping in order to have lower OFDM sidelobe. The original symbol after M-PSK mapping which is located at point *a* of *M* available PSK symbol points is transformed into two possible new points a_1 and a_2, where a_1=-a_2. By combining all possible new OFDM symbol set on the 2x M-PSK points through all OFDM carriers, the one set which gives the smallest sidelobes will be chosen for the transmission. No additional information is required by the receiver for data detection, since the receiver has the knowledge of all possible a_1 and a_2 positions. Therefore, the receiver can detect the data by calculating the minimum Euclidean distance between the received symbol and the reference points on 2xM- PSK mapping, and accordingly translates the result to the M PSK mapping. Figure 5 shows an example on how the points from QPSK are expanded to 8-PSK mapping. In this way there are two possibilities of mapping a data symbol, and accordingly the constellation expansion will add extra randomness to X_n in (1) with various possible sidelobes magnitude. The one which gives the smallest sidelobes will be chosen to be transmitted. The simulation results in [24] have shown that the sidelobes suppression technique by the constellation expansion from QPSK to 8-PSK gives only slight BER degradation. With Constellation Expansion technique about 9 dB sidelobes suppression can be attained [24].

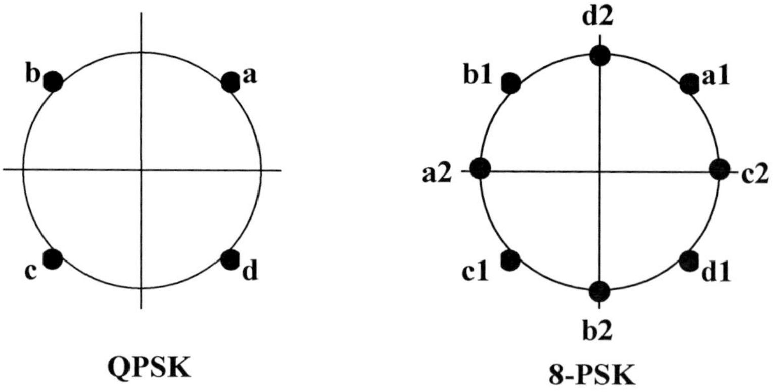

Figure 5. Constellation Expansion from QPSK to 8-PSK mapping.

The work has been extended in [25] by combining the constellation expansion technique with the cancellation carriers inserted on the carriers adjacent to the LU band. Extra sidelobes reduction has been achieved without having high value of cancellation carrers magnitude, due to prior sidelobes reduction by the constellation expansion technique. According to [25] the combination between constellation expansion and one sidelobes cancellation carrier on each side adjacent to LU band gives around 16 dB sidelobes reduction.

2.2. Wavelet Packet Based Cognitive Radio System

In multicarrier wavelet packet modulation (MC-WPM), the Fourier transform used in OFDM is replaced by the orthonormal wavelet packet bases. The wavelet packet bases and their dual bases are obtained from perfect reconstruction two-band FIR filter bank solution using multistage tree-structured paraunitary filter banks derived by cascading 2-channel filter banks [26][27]. The wavelet transform provides higher sidelobes suppression when compared to the rectangular window sidelobes in regular OFDM [28]. In [29] the Daubechies and Haar wavelet-based multicarrier transmission over Rayleigh fading channel were evaluated. Their performances in terms of intersymbol interference (ISI) and intercarrier interference (ICI) have been compared with the Fourier-based OFDM using rectangular window. The wavelet-based multicarrier provides significant ICI power reduction and slight ISI power reduction compared to OFDM.

2.2.1. Wavelet Packet Filterbanks

The wavelet packet subcarriers (to be used at the transmitter end) are generated through a multichannel filterbank consisting of two cascaded channel filterbanks in which the synthesis filters (H' and G') are applied. The process known as the synthesis procedure consists of binary interpolation (upsampling) by 2, filtering and recombination at each level. To demonstrate the generation of the wavelet packet bases process, we shall consider here a cascaded level-3 tree structure of the type shown in Figure 6(a). Such an arrangement can give rise to eight wavelet packet bases. In general, a level-J tree has to be constructed to generate M bases or sub-carrier waveforms. The structure in Figure 6(a) can be simplified by applying the noble identities procedure that will produce multi rate systems. Each of the carrier will be upsampled by M and will experience only one tap filtering. The filter is the equivalent filter resulting from the convolution of the combination between low pass and high pass filters according to the branch line of the tree [30]. The high pass filter $H'(z)$ with length L' can be obtained from the low pass filter $G'(z)$ by means of the $G'(z)=-z^{-(L'-1)}H'(-z^{-1})$ equation [31]. The coexistence with the LUs is actualized by shaping its transmission waveform and by adaptively activating or deactivating subcarriers in such a way as to utilize the unoccupied time-frequency gaps of the LU. The idea is to dynamically notch the transmission signal so that it has no or very little time-frequency components competing with the LU. In that way the CR can seamlessly blend with the LU operation.

The wavelet packet duals (to be used at the receiver end) are also obtained from multichannel filter bank analysis, though the processes are reversed. The duals are obtained from the analysis filters (*H* and *G*) through the analysis procedure which consists of filtering, decimation (downsampling) by 2 and decomposition at each stage. Figure 6(b) illustrates the generation of 8 wavelet packet duals from a level-3 tree cascaded filter bank. The dual filters are derived from paraunitary condition $P(z) + P(-z) = 2$, where the product filter $P(z) = H(z)H(z^{-1})$, [31] and $H(z^{-1}) = H'(z)$. In the time domain the analysis filters are the complex conjugate time reversed versions of the synthesis filters ($h'(n) = h^*(-n)$ and $g'(n) = g^*(-n)$).

The wavelet basis function in the application of MC-WPM systems should fulfill some fundamental properties. These properties include the time and band limitation, the orthogonality between the bases and their duals for perfect reconstruction, and the orthogonality between the bases for the unique demodulation of each basis function. Of the several available wavelets such as: Coifflets, Daubechies, Haar, Symlets, the most suitable wavelet that fulfils the required properties is the family of maximum frequency selective filter banks derived from a modified Remez exchange algorithm [31]. It provides a degree of freedom to shape its sidelobes by varying its regularity order, transition bandwidth and filter length. From the filter with length L', the Remez algorithm will produce L' equations, $L'/2$ of which will be dedicated for the paraunitary condition, K equations for the regularity order and left $L'/2 - K$ for frequency selectivity. The longer the filter length, the greater the degree of freedom will be for the design of the frequency selectivity and the regularity order of the filters. The smaller the transition bandwidth, the more frequency selective the filter will be.

Regularity corresponds to the smoothness and sidelobes of the signal. As the regularity order increases, so the sidelobes of the signal become lower and smoother. There will be trade off between CR performance and sidelobe reduction. The sidelobe can be depleted by increasing the regularity order with the cost of low filter frequency selectivity that will deteriorate the CR performance. More details on the analytical aspect of this subject can be found in [31]. The performance of MC-WPM with Remez exchange algorithm as CR system is evaluated in [32], and is compared with the Fourier-based MC-OFDM CR system in the presence of the LU signal in the Additive White Gaussian Noise (AWGN) channel. The results show that within certain Remez parameter values, MC-WPM outperforms the Fourier-based MC-OFDM.

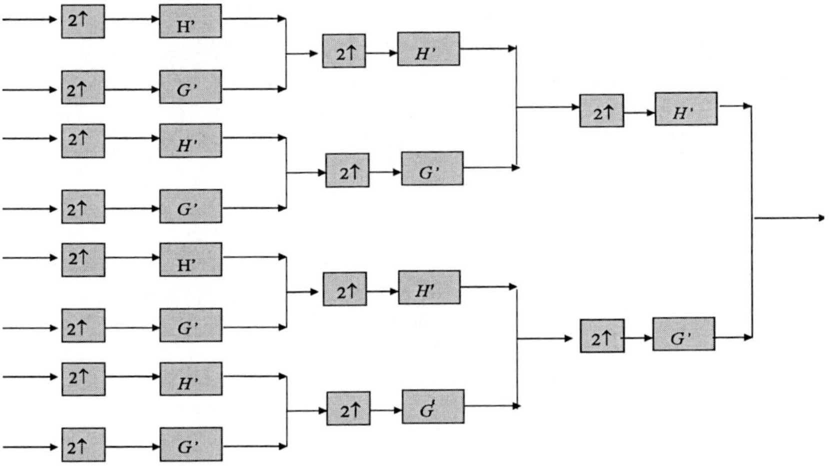

(a)

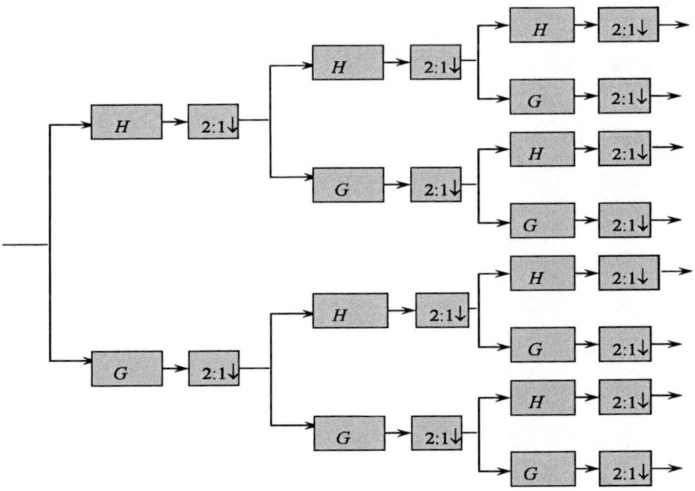

(b)

Figure 6. Generation of wavelets. (a) A level-3 tree gives 8 wavelet packet bases for the transmitter and their (b) duals for the receiver. The up arrows represent interpolation by 2, the down arrows represent decimation by 2. *H'* and *G'*, denote the frequency responses of the low and high pass reconstruction filters, while *H* and *G* denote the frequency responses of the low and high pass decomposition filters, respectively.

2.2.2. Interpolated Tree Orthogonal Multiplexing (ITOM)

In the MC-WPM technique the filter banks perform the dual role of shaping the spectrum and also interpolating in time series. A more intelligent method is to separate the two processes in order to gain a greater control over the characteristics of the carriers. This method, called the Interpolated Tree Orthogonal Multiplexing (ITOM) method, was introduced in [13] . The spectral shaping is performed in a bank of shaping filters external to the wavelet packet tree structure. The procedure is depicted in Figure 7 where we can see that up-sampled shaping filters precede the input ports of the wavelet packet tree structure.

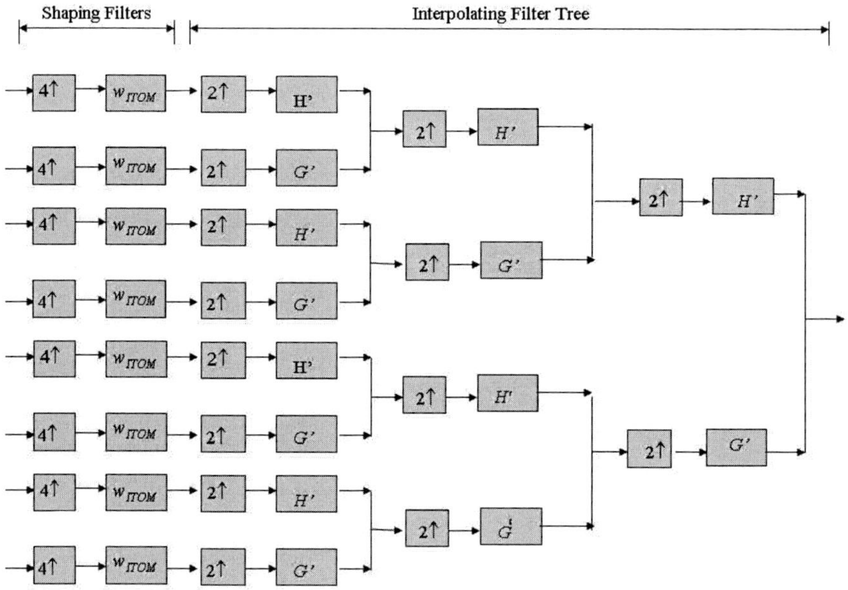

Figure 7. Modulator of ITOM Interpolated Tree Structure [13] where w_{ITOM} is the ITOM shaping filters.

The prototype baseband spectral shape is preserved during the interpolation and aliasing operations. Notching over the desired spectral interval is achieved by vacating one or more of the input branches. In principal the ITOM will deliver the spectra formed by the input shaping filters to specific center frequencies by sequentially forming aliased spectral replicas by up sampling and selective half-band filtering the desired replica [13].

3. Single Carrier Cognitive Radio System

Single carrier transmission for opportunistic spectrum access has recently being initiated. Upon receiving the spectrum occupancy information the time domain signal can be shaped in a way that part of the spectrum will not distort the legacy system signal. In the next sections single carrier transmission using Fourier and Wavelet transformation signal shaping techniques will be discussed.

3.1. Transform Domain Communications System (TDCS)

Transform domain communication system (TDCS) is a single carrier modulation that was initially introduced in [33]. TDCS fundamentally synthesizes a smart adaptive waveform to avoid interference at the transmitter, while the conventional technique tries to mitigate the interference at the receiver. Basically the spectral of interference signal or the LU's signal currently occupying the spectrum are scanned. Once they are indentified through some form of threshold detection, a clean spectrum information $\boldsymbol{A}= \{A_1, A_2, ..., A_{NFFT}\}$ will be formed by the spectrum magnitude module, where amplitudes of interfering frequency components exceeding a threshold are set at zero, while the others are set at one. The multi-valued complex pseudorandom phase vector ($e^{j\boldsymbol{\theta}}$) is multiplied element by element with $\boldsymbol{A}$ to produce the spectral vector $\boldsymbol{B}$. After IFFT operation, a time domain basis function $b(t)$ is derived, which is subsequently stored and modulated by using pulse amplitude modulation (PAM), pulse position modulation (PPM) or cyclic code shift keying (CCSK). In CCSK the bits are represented by the cyclic shifted pulse in time.

This zeros insertion impact can be described from the TDCS power spectral density (PSD),

$$PSD(f) = \left| \frac{1}{N_{FFT}} \sum_{m=0}^{N_{FFT}-1} A_m e^{j\theta_m} \int_{-(1+\alpha)\frac{T_u}{2}}^{(1+\alpha)\frac{T_u}{2}} p(t) e^{-j2\pi(f-f_m)t} dt \right|^2 \tag{9}$$

where T_u is the useful signal duration, $p(t)$ is the window function, α is the roll off factor of the window, and θ_m the phase on subband m produced by the random phase module. The random phase can be derived from one of the 2^r phase shifts which are evenly spaced around a unit circle, while r is the number of bits representing a complex point if a linear feedback shift registers is used to choose it. The primitive polynomial of the linear feedback shift registers can be configured for multiple access features so that each access has its own random phase sequence. The random phase generator should be designed in such a way that the resulting time domain basis function $b(t)$ has correlation properties similar to those of sampled noise. If $p(t)$ is a rectangular window and α is zero, the area within the integral can be replaced by $T_u sinc((f - f_m) T_u)$. Zero amplitude at the carrier m will make the PSD on that carrier position zero, and due to the orthogonality among carriers the power contributions from the other carriers are also zero. In this way the TDCS can be used as CR modulation technique in overlay opportunistic spectrum access, since part of its spectrum on the LU band will not significantly impact the LU transmission. Equation (9) is almost similar to (1), only the power p_m and symbol amplitude X_m in (1) are replaced by the spectrum information A_m and the random phase θ_m.

TDCS gives greater degree of freedom in choosing the window $p(t)$ to lower the sidelobes of its spectrum. Unlike OFDM, the spectrum sidelobes of the window for TDCS do not have to be zero on the frequency spacing interval (in OFDM this is described as the orthogonality between carriers), as long as its sidelobes are very small. Since data detection at the receiver is performed by correlating the received signal with the reference signal, the designed non-orthogonal spectrum e.g. by applying half sine window or Gaussian window to the transmitted signal will not affect data detection. De-correlation for symbol demapping is simplified in [34] by choosing the index (position) of the maximum of absolute value of inverse Fourier transform of the product of the conjugate received signal and reference signal in frequency domain.

3.1.1. TDCS with an Extra Embedded Symbol

Recently the modified the TDCS model is proposed by adding another bits source as a form of embedded symbol and by multiplying the frequency domain TDCS basis function with the embedded symbol [35]. The purpose is to increase the TDCS bitrate. This extra bits source can also be used to transmit CR side information. Assigning extra bits to the embedded symbol denotes allocating extra power to the TDCS basis function (increased SNR per symbol) that will lead to the data detection enhancement of the legacy data source (from PAM, PPM or CCSK detection). The TDCS with extra embedded symbol transmitter and receiver schemes are depicted in Figure 8.

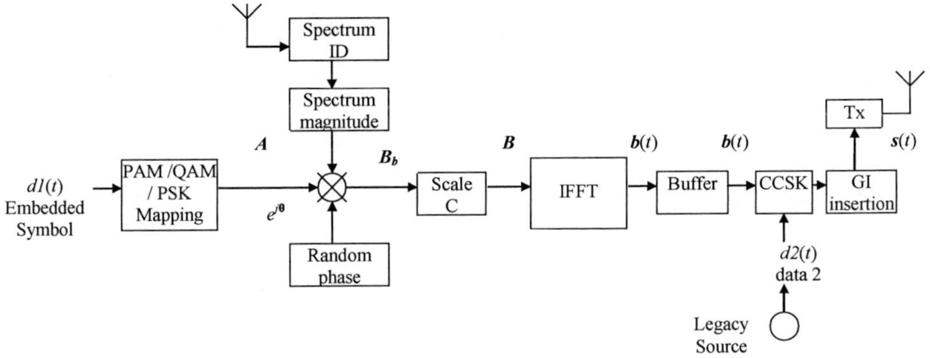

(a)

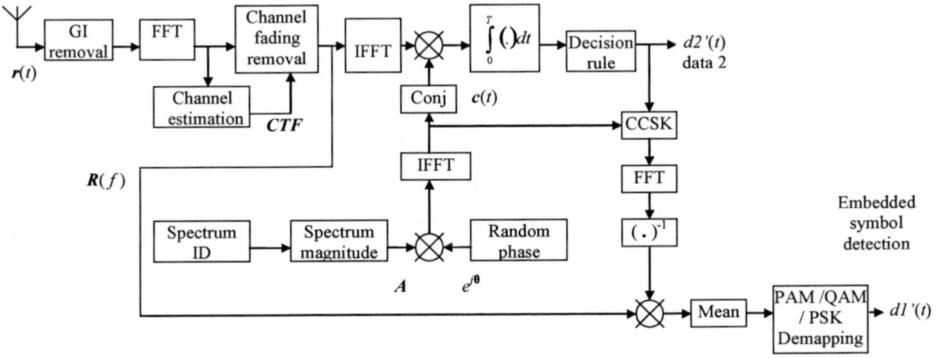

(b)

Figure 8. (a) Transmitter and (b) receiver architecture of TDCS with extra embedded symbol featuring channel estimation module.

Due to the extra embedded symbol, the TDCS PSD would become :

$$PSD(f) = \left| \frac{1}{N_{FFT}} S e^{j\varphi} \sum_{m=0}^{N_{FFT}-1} A_m e^{j\theta_m} \int_{-(1+\alpha)\frac{T_u}{2}}^{(1+\alpha)\frac{T_u}{2}} p(t) e^{-j2\pi(f-f_m)t} dt \right|^2 \tag{10}$$

where S and φ are the embedded symbol magnitude and phase. In order not to alter the PSD amplitude the value of S should be 1, therefore phase shift keying (PSK) is considered as the suitable embedded symbol modulation mode.

3.2. Wavelet Domain Communications System (WDCS)

Another family of TDCS is the wavelet domain communication system (WDCS) [29][30] where the Fourier transform operation in TDCS is replaced by wavelet transformation. The wavelet transform is used in this scheme to identify and establish an interference-free spectrum. It was initially performed to overcome the deficiencies of the TDCS in estimating the characteristics of non stationary interference, and the inefficiency of the Fourier-based spectral estimator causing spreading of the interference energy into adjacent spectral regions thus resulting in performance degradation [36]. In [37] the traditional wavelet transform is extended by introducing the wavelet packet decomposition technique to WDCS with the intention of increasing the transform adaptation over a larger class of interfering signals, to obtain finer high-frequency resolution, and more accurate electromagnetic spectral estimates. Through the use of adaptive thresholds and notches, subbands in the packet-based WDCS transform that contain the interference are effectively cancelled.

A WDCS time domain basis function $b(t)$ is obtained by performing an inverse wavelet transfom of the multiplication (point to point) between the spectrum occupancy vector $\boldsymbol{A}$ and the random phase $e^{j\boldsymbol{\theta}}$. The wavelet basis functions which are earlier applied in MC-WPM such as Coifflets, Daubechies, Haar, Symlets or Remez filter can also be used in WDCS. The time domain basis function $b(t)$ is later modulated with the data using PAM, or PPM, or CCSK modulation and then transmitted. The WDCS transmission block is presented in Figure 9.

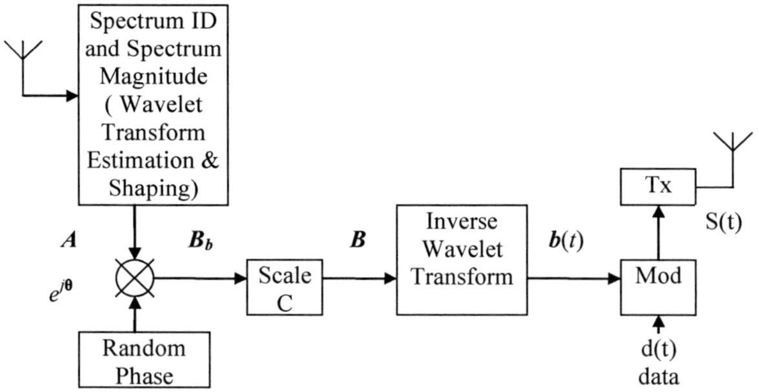

Figure 9. WDCS Transmitter blocks [29].

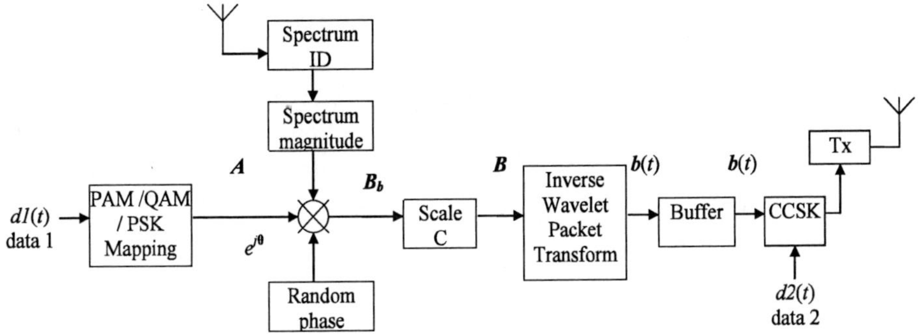

(a)

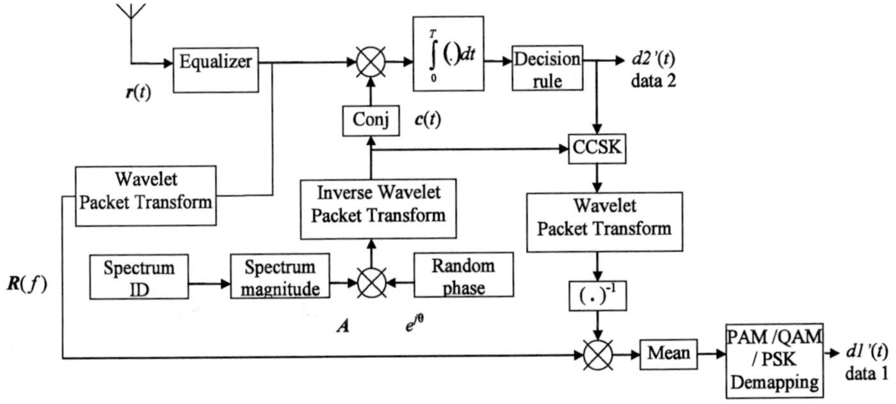

(b)

Figure 10. (a) Transmitter and (b) Receiver architecture of WDCS with extra embedded symbol.

At the receiver the detection is preceded by correlating the received signal with the reference signal in time domain. The receiver structure of WDCS is virtually the same as the TDCS, only the FFT and IFFT modules are replaced by respective discrete wavelet transform (DWT) and inverse discrete wavelet transform (IDWT).

3.2.2. WDCS with an Extra Embedded Symbol

The application of the embedded symbol to WDCS is presented in Figure 10. According to [28] the impact of fading in the modulation technique using the wavelet basis function would be properly removed by time-domain equalization. Therefore, different with the TDCS scheme (Figure 8) the frequency-domain channel estimation and fading removal is replaced by time-domain equalization.

Further, the CCSK detection in WDCS with extra embedded symbol system is applied in time-domain, while the embedded symbol detection remains conducted in frequency-domain. The frequency domain embedded symbol detection is done due to the interference avoidance reason. By neglecting the signal at the LU band position (Spectrum Pooling technique), the embedded symbol will not be affected by the interference from the LU signal.

4. Summary

By relying on the accurate information from the spectrum sensing module about spectrum occupancy, a cognitive radio system can shape its signal in such a way that it coexists with the neighboring licensed users and tries to achieve reliable communications while accessing the spectrum opportunistically. Of the several modulation techniques, it is the OFDM that is considered to be a strong candidate due to its spectrum access flexibility by notching its carriers on the position of licensed user's band. The main OFDM drawback is that its high sidelobes will interfere with the neighboring licensed user's band. Techniques such as windowing, adjacent carriers deactivation, cancellation carriers, multiple choice sequence and a combination of all these techniques as the solutions used to mitigate the sidelobes problem have been overviewed. While windowing technique is a known application for sidelobes suppression, these techniques discussed in this paper are novel for the CR. Bit rate loss due to windowing can be compensated by the application of adaptive bit loading, where more bits are transmitted on the carriers with good SNR and fewer bits are transmitted on the carriers with bad SNR, while the target rate still can be obtained. Moreover, the OFDM symbols can be overlapped with the deliberate design of overlap position that will prevent the inter symbol interference.

The Fourier transform in OFDM can be replaced by taking the wavelet packet based transform as an alternative technique. References have shown that the wavelet packet based technique provides lower spectrum sidelobes compared to the regular OFDM with rectangular window. Among several wavelet basis functions, it is the one which is derived from the Remez exchange algorithm that provides a higher degree of flexibility in shaping its signal spectrum. By configuring the filter length, transition bandwidth and regularity order of the filter, the desired cognitive radio spectrum can be designed.

Single carrier modulation for a cognitive radio system known as transform domain communications system (TDCS) has been studied. The technique requires the spectrum occupancy information and relies for its data detection on the correlation properties of the generated random sequence used to construct its waveform. By utilizing the spectrum occupancy information and the random sequence, the TDCS can shape its signal to share the spectrum with the licensed users. The shaping is employed by notching its energy in the band occupied by the licensed user. The combination with windowing will even shape its spectrum better. Since the data detection at the receiver is applied by cross-correlating the received signal with the reference signal, subcarrier unorthogonality will not severely affect the TDCS data detection. This condition indicates a huge degree of freedom when it comes to choose the window shape to lower the TDCS spectrum sidelobes. The data bits are aided to the signal by PAM, PPM, or CCSK modulation. Like OFDM, TDCS also requires an IFFT module to create its time domain signal. TDCS possesses lower bitrate compared to OFDM. As an effort to increase the TDCS bitrate we have introduced additional data source to the TDCS as a form of embedded symbol by multiplying the frequency domain TDCS basis function with the embedded symbol. As embedding extra symbol to the TDCS basis function refers to adding extra energy to the TDCS symbol, the cross-correlation property of the TDCS symbol will be enhanced. As the outcome the data BER from the legacy bit source (PAM, PPM or CCSK) will be improved.

The Fourier transform in TDCS can be replaced by the wavelet transform that invokes the WDCS as another single carrier modulation technique for cognitive radio system. The data detection in WDCS is implemented by correlating the received WDCS signal with the time domain reference signal.

Acknowledgment

The authors acknowledge the support of Dutch Adaptive Adhoc Freeband (AAF) project in this research, and the contribution from Madan Kumar Lakshmanan related to the wavelet packet multicarrier modulation topic. This chapter is dedicated to the memory of the first author father, Baik Ukurta Purba.

References

[1] Haykin, S. (2005). "Cognitive Radio: Brain Empowered Wireless Communications", in *IEEE Journal on Selected Areas In Communications*, vol. 23, no.2, Feb. 2005, 201-220.

[2] Joseph Mitola III, *Cognitive Radio An Integrated Agent Architecture for Software Defined Radio*, PhD thesis, Dept. of Teleinformatics, KTH Royal

Institute of Technology, Stockholm, Sweden, 2000.

[3] Brodersen, R. W., Wolisz, A., Cabric, D., Mishra, S. M. & Wilkomm, D. (2004). "CORVUS : A Cognitive Radio Approach for Usage of Virtual Unlicensed Spectrum. White Paper, Dept of electrical Engineering and Computer Science University of Berkeley, July, 2004. http://bwrc.eecs.berkeley.edu/Research/ MCMA/CR_White_paper_final1.pdf.

[4] Weiss, T., Hillenbrand, J., Krohn, A. & Jondral, F. K. (2004). "Mutual Interference in OFDM-based Spectrum Pooling Systems", in *Proc.IEEE Vehicular Technology Conference*, Milan, Italy, 17-19 May 2004, 1873-1877.

[5] Weiss, T. & Jondral, F. (2004). "Spectrum Pooling - An Innovative Strategy for the Enhancement of Spectrum Efficiency", in *IEEE Communications Magazine*, vol. 42, no. 3, March 2004, S8-14.

[6] Nikookar, H. & Prasad, R. (1997). "Wave Shaping of Multicarrier Signal for Data Transmission Over Wireless Channels", in *6 th IEEE International Conference on Universal Personal Communications Record (ICUPC)*, 173-177.

[7] Tan, P. & Beaulieu, N. C. (2004). "Reduced ICI in OFDM Systems Using the Better than Raised – Cosine Pulse". *in IEEE Communications Letter*, vol. 8, no. 3, March 2004, 135-137.

[8] Assalini, A. & Tonello, A. M. (2004). "Improved Nyquist Pulses", in *IEEE Communications Letters*, vol. 8, no.2, 87-89.

[9] Muschalik, C. (1996). "Improving An OFDM Reception Using An Adaptive Nyquist Windowing", in *IEEE Transactions on Consumer Electronics*, vol. 42, no.3, 259-269.

[10] Witrisal, K. (2002). "*OFDM Air Interface Design for Multimedia Communications*", Ph.D. Thesis, Dept. Electrical Eng. Mathematics and Computer Science, Delft University of Technology, Apr. 2002.

[11] Budiarjo, I., Nikookar, H. & Ligthart, L. P. (2006). "Combined Spectrum Pooling and Adaptive Bit Loading for Cognitive Radio OFDM Based System", in *Proc. 13th IEEE SCVT*, Liege, Belgium, Nov. 2006, 73-76.

[12] Nikookar, H. & Prasad, R. (1997). "Optimal Waveform Design for Multicarrier Transmission Over a Multipath Channel", in *Proc. 47th IEEE Vehicular Technology Conference*, Phoenix, Arizona, USA, May 4-7, 1812-1816.

[13] Harris, F. & Kjeldsen, E. (2006). "A Novel Interpolated Tree Orthogonal Multiplexing (ITOM) Scheme with Compact Time-Frequency Localization: an Introduction and Comparison to Wavelet Filter Banks and Polyphase

Filter Banks", in *Software Defined RadioTechnical Conference and Product Exposition*, Orlando, Florida, USA, Nov. 2006.

[14] Yamaguchi, H. (2004). "Active Interference Cancellation Technique for MB-OFDM Cognitive Radio", *Proc. 34 th European Microwave Conference*, Amsterdam, The Netherlands, vol. 2, Oct. 2004, 1105-1108.

[15] Brandes, S., Cosovic, I. & Schnell, M. (2005). "Sidelobe Suppression in OFDM Systems by Insertion of Cancellation Carriers", *Proc. 62nd IEEE Vehicular Technology Conference (VTC)*, Fall, Dallas, USA, Sept, 2005.

[16] Brandes, S., Cosovic, I. & Schnell, M. (2006). "Reduction of Out of Band Radiation in OFDM Systems by Insertion of Cancellation Carriers", in *IEEE Comunications Letters*, vol. 10, issue 6, June 2006, 420-422.

[17] Brandes, S., Cosovic, I. & Schnell, M. (2005). "Reduction of Out-of-Band Radiation in OFDM Based Overlay Systems", in *Proc. First IEEE International Symposium New Frontiers in Dynamic Spectrum Access Networks (DySPAN)*, Baltimore, Maryland, USA, Nov. 2005, 662-665.

[18] Berthold, U., Jondral, F. K., Brandes, S. & Schnell, M. (2007). "OFDM Based Overlay Systems: A Promising Approach for Enhancing Spectral Efficiency" in *IEEE Communications Magazine*, vol. 45, issue 12, Dec. 2007, 52-58.

[19] Cosovic, I., Brandes, S. & Schnell, M. (2006). "Subcarrier Weighting : A Method for Sidelobe Suppression in OFDM Systems", in *IEEE Communications Letters*, vol. 10, no. 6, June 2006, 444-446.

[20] Nikookar, H. & Prasad, R. (2000). "Weighted OFDM for Wireless Multipath Channels", in *IEICE Trans. on Commun.*, vol. E83-B, no. 8, Aug. 2000, 1864-1872.

[21] Cosovic, I. & Mazzoni, T. (2006). "Suppression of Sidelobes in OFDM systems by multiple choice sequences", *in European Transactions on Telecommunications Special Issue on Multi carrier – Spread Spectrum Wiley InterScience*, June 2006, 623-630.

[22] Mahmoud, H. A. & Arslan, H. (2008). "Sidelobe Suppression in OFDM – Based Spectrum Sharing Systems by Using Adaptive Symbol Transition", in *IEEE Communications Letters*, vol. 12, no. 2, Feb. 2008, 133-135.

[23] Cosovic, I. & Mazzoni, T. (2007). "Sidelobe Suppression in OFDM Spectrum Sharing Sys-tems via Additive Signal Method", in *Proc. 65th IEEE Vehicular Technology Conference (VTC)*, Dublin, Ireland, Apr. 2007, 2692-2696.

[24] Pagadarai, S., Rajbanshi, R., Wyglinski, A. M. & Minden, G. J. (2008). "Sidelobe Suppression for OFDM-Based Cognitive Radios Using Constellation Expansion", in *Proc. IEEE Wireless Communicaitons and*

Networking Conference (WCNC), Las Vegas, Nevada, USA, Apr. 2008, 888-893.

[25] Pagadarai, S., Wyglinski, A. M. & Rajbanshi, R. (2008). "A Novel Sidelobe Suppression Technique for OFDM-Based Cognitive Radio Transmission", in Proc. *IEEE Symposia on New Frontiers Dynamic Spectrum Access Networks (DySPAN)*, Chicago, USA, Oct. 2008.

[26] Vetterli, M. & Kovacevic, I. (1995). *Wavelets and Subband Coding*, Prentice Hall PTR, Engle-wood Cliffs, New Jersey.

[27] Jamin, A. & Mahonen, P. (2005). "Wavelet Packet Modulation for Wireless Communications", in *Wireless Communications & Mobile Computing Journal, vol. 5, issue 2*, John Wiley and Sons Ltd., March 2005, 123-137.

[28] Bouwel, C. V., Potemans, J., Schepers, S., Nauwelaers, B. & Capelle, A. V. (2000). "Wavelet Packet Based Multicarrier Modulation", in *Proc. IEEE Benelux Symposium on Communications and Vehicular Technology*, Leuven, Belgium, Oct. 2000, 131-138.

[29] Negash, B. G. & Nikookar, H. (2000). "Wavelet Based Multicarrier Transmission Over Multipath Wireless Channels", in *IEE Electronics Letters*, vol. 36, no. 21, Oct. 2000, 1787-1788.

[30] Zhang, Y. & Cheng, S. (2004). "A Novel Multicarrier Signal Transmission System Over Multi-path Channel of Low-Voltage Power Line", in *IEEE Transactions on Power Delivery*, vol. 19, no. 4, Oct. 2004, 1668-1672.

[31] Rioul, O. & Duhamel, P. (1994). "A Remez Exchange Algorithm for Orthonormal Wavelets", in *IEEE Trans. Circuits Systems - II*, vol. 41, no. 8, 550-560, Aug. 1994.

[32] Lakshmanan, M. K., Budiarjo, I. & Nikookar, H. (2007). "Maximally Frequency Selective Wavelet Packets Based Multi-Carrier Modulation Scheme for Cognitive Radio Systems", in *Proc. 50'th IEEE Global Communications Conference (Globecom)*, Washingthon DC, USA, Nov. 2007, 4185-4189.

[33] Chakravarthy, V., Nunez, A. S., Stephens, J. P., Shaw, A. K. & Temple, M. A. (2005). "TDCS, OFDM, and MC-CDMA: A brief tutorial," *IEEE Communications Magazine*, vol. 43, no.9, Sept. 2005, S11-S16.

[34] Dillard, G. M., Reuter, M., Zeiddler, J. & Zeidler, B. (2003). "Cyclic Code Shift Keying: A Low Probability of Intercept Communication Technique", in *IEEE Trans. on Aerospace and Electronic Systems*, vol.39, no.3, July 2003, 786-798.

[35] Budiarjo, I., Nikookar, H. & Ligthart, L. P. (2008). "On the Utilization of Embedded Symbol for CCSK BER Improvement in TDCS Dynamic Spectrum Access", in *Proc. IEEE European Wireless Technology*

Conference, 28-29 Oct 2008, Amsterdam, The Netherlands.

[36] Lee, M. J., Temple, M. A., Claypoole, R. L. Jr., Raines, R. A. & Stephens, J. P. (2002). "Wavelet Domain Communication System: Bit Error Sensitivity Characterization for Geographically Separated Transceivers", in *Proc. MILCOM*, Anaheim, CA, Oct. 2002, vol. 2, 1378-82.

[37] Lee, M. J. (2001). "*Wavelet Domain Communication System (WDCS): Packet- Based Wavelet Spectral Estimation and M-ARY Signaling*", Master thesis, AFIT/GE/ENG/02M-14, Graduate School of Engineering and Management Air Force Institute of Technology Wright – Patterson Air Force Base, Ohio, USA, Mar. 2001, DTIC:ADA401433, approved for public release.

In: Cognitive Radio
Editors: P.D. McGuire et al, pp. 101-120
ISBN: 978-1-60876-604-8

Chapter 4

WHITE SPACE COGNITIVE RADIO COMMUNICATION SYSTEM COGSIP AND RELATED TECHNIQUES

Tazuko Tomioka, Ren Sakata, Koichiro Ban and Takahiro Kobayashi
Corporate Research & Development Center, Toshiba Corp.

Abstract

Studies of white-space cognitive radio systems are being actively pursued in response to the frequency depletion problem. At present, a signal of lumped frequency is placed in one of the white-space frequency bands, such as TV bands. In the future when cognitive radio techniques are mature, many frequency bands with discrete center frequencies may be used simultaneously in order to transmit extremely broadband data. However, many problems have to be solved in order to realize the system. In view of this prospect, we have proposed a broadband short-range white-space cognitive system COGSIP (COGnitive System of Intelligent Processing) to substantiate the problems and investigate them. COGSIP is a personal area network (PAN) system with a data rate of a few Gbps in the microwave band. A COGSIP terminal searches white spaces from the broad microwave band and utilizes many white spaces simultaneously avoiding primary transmissions to realize a data rate of a few Gbps. In this chapter, we introduce the outline of COGSIP and four related techniques we have investigated: high-sensitivity energy detection, narrow-band signal detection, initial broadcasting signal and transmission signal shaping. Since COGSIP is a PAN system, it uses energy detection owing to its low processing load, and we introduce method of sensitivity increased by overlapped FFT. Next, a detection method of narrow band signals using autocorrelation is introduced. It achieves detection of a narrow-band signal whose bandwidth is much smaller than an FFT bin width and which cannot be detected by energy detection. Then, an initial broadcasting signal scheme is introduced, which aims to realize announcement of control information to a newly joining terminal that does not know the frequency utilized. The broadcasting is realized without any fixed-frequency control channel. Finally, a transmission signal shaping technique to realize low PAPR (peak-to-average power ratio) in order to reduce interference on primary signals between the COGSIP utilizing bands is introduced. The technique is a modification of DFT-spread OFDM.

1. Introduction

Studies of cognitive radio are being actively pursued in response to the frequency depletion problem [1]-[3]. In the U.S., the FCC (Federal Communications Commission) has decided to open up the TV white spaces that will become available through rearrangement of VHF (very high frequency) and UHF (ultra high frequency) bands accompanying the transition from analogue broadcasting to digital broadcasting, to cognitive radio communication systems [4]. Also, Ofcom (Office of Communications) in the U.K. expressed the opinion that it would be desirable for frequency spaces available after digital transition to be used as white spaces, which is an approach similar to that being pursued in the U.S. [5]. Thus, white spaces are expected to be used in next-generation wireless systems. As for standardization, in IEEE 802.22 [6], a cognitive WRAN (wireless regional area network) system used in TV white spaces is being standardized and a study group for developing a mobile white space standard has been setup [7]. Moreover, the concept of frequency dynamic auction has been proposed [8]. In this way, very flexible frequency utilization is anticipated in the future. In the foreseeable future, the wireless interface of a white space cognitive radio terminal will be very similar to conventional wireless interface, which uses only one band at a time except for carrier-sensing function. The bandwidth of a band is expected to be from a few MHz to a few tens of MHz. For example, the wireless interface of IEEE 802.22 is very similar to that of IEEE 802.16 except that it has a white space detection function.

Looking further ahead, very wide white spaces other than TV white spaces may be available, allowing wider total bandwidth to be used in a very small area, such as in users' houses or rooms. In such cases, high rate and stable transmission will be possible if many discrete white spaces are used at once. This method can make stable Gbps transmission possible in the future. However, the terminal will be absolutely secondary, and when primary transmission is detected in a frequency band, the cognitive terminal should stop the transmission in the detected frequency band. The authors have assumed such usage and have proposed a virtual system design of a short-range high-rate cognitive system called COGSIP* (COGnitive System of Intelligent Processing) [9] to clarify and investigate the problems of a system in which many discrete bands are simultaneously used, with primary system bands, which should be kept free from interference, existing between the bands. In this chapter, the outline of COGSIP and several investigated techniques for COGSIP are introduced.

2. The Outline of Cogsip

COGSIP is a personal area network (PAN) system for indoor use. As indicated in Figure 1, the target applications are indoor video transmission or high-speed file transfer. The terminals use frequencies such that primary transmission from inside and outside of the room is avoided. Table 1 shows the target specification of COGSIP. The maximum communication distance is 30 m and somewhat exceeds the range of PAN. The range and data rate trade-off and the maximum range will be achieved only with short range and fine propagation environment. Mobility is low because of indoor use. The system frequency band is somewhat low microwave band in order to achieve non-line-of-sight (NLOS) communication.

As indicated in Figure 2, a terminal selects one or two bands, called "main bands", from the system frequency band and places "sub-bands" to perform transmission at the available white spaces in the main band. A terminal selects a main band frequency according to the conditions such as communication distance, application, and/or required data rate and it selects sub-band frequencies according to the environmental frequency utilization condition. In order to obtain the environmental frequency utilization information, the terminal performs carrier sensing frequently. Moreover, there will be frequency masks defined by regulations for cognitive systems and therefore a terminal excludes the masked frequency from the sub-band selection. If there is a data-base listing available white spaces of each time and place, the terminal obtains the information of available frequencies at the time and at the place and decides the main bands and sub-bands to be utilized.

* "COGSIP" is a trademark of Toshiba Corporation.

The terminal converts the whole main band from analogue to digital or vice versa using an analogue-to-digital converter (ADC) and a digital-to-analogue converter (DAC) and processes all the signals inside a main band at once by the digital processing unit.

Table 1. Target specification of COGSIP.

Items	Values
Communication Distance	30 m max.
Transmission Data Rate	4.5Gbps max.
Mobility	Static, pedestrian
Topology	1:1, 1:n, n:n. Ad hoc
Environment	Indoors
System frequency band	3 GHz–12 GHz
Main band bandwidth	1 GHz
Maximum number of main bands	2 each for TX and RX

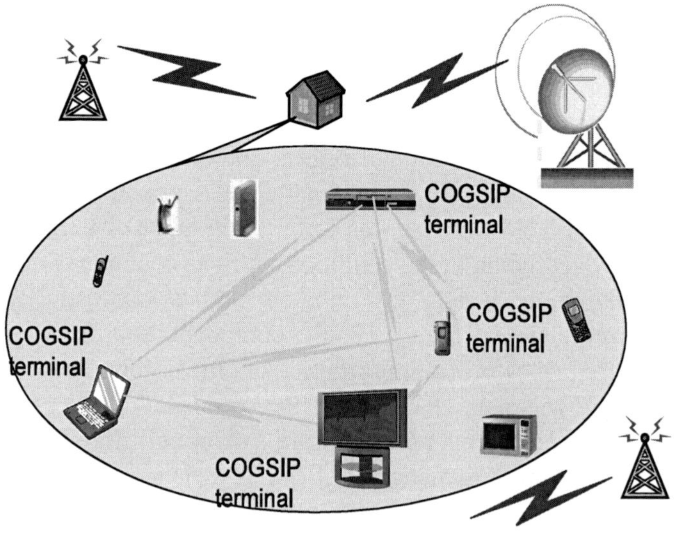

Figure 1. Usage of COGSIP.

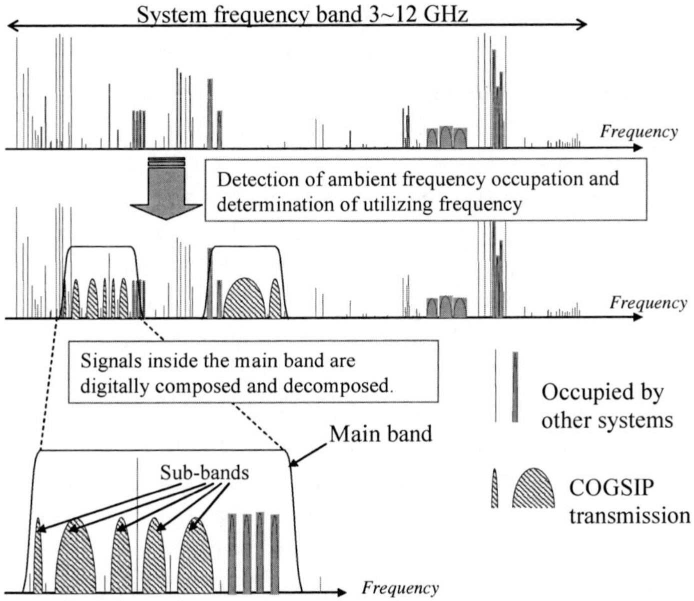

Figure 2. Frequency Utilization.

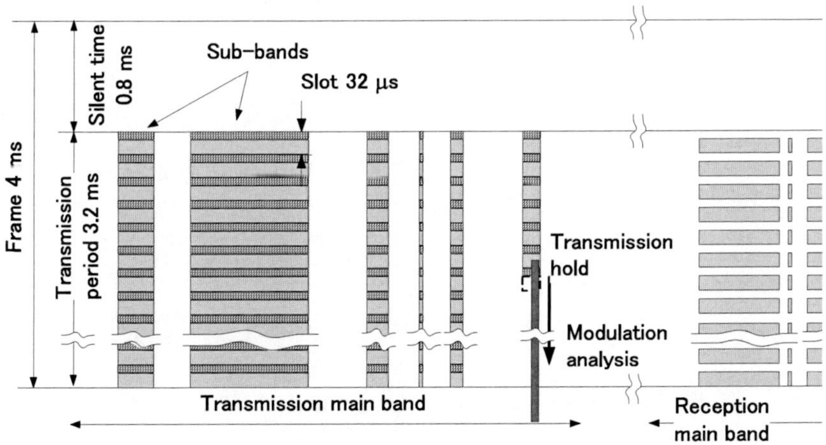

Figure 3. Time usage of a main band.

Figure 3 shows frame and slot usage of the system. The horizontal axis is frequency and the vertical axis is time. In order to detect primary signal securely, the terminal performs carrier sensing very frequently. A frame consists of 100 slots and 0.8 ms silent time, and the frame length is 4 ms. In the silent time, the terminal does not transmit any signal. The silent time has two purposes: detection of the detailed frequency utilization for the main band and making a vacant time so as not to interfere the carrier sensing of primary terminals. Therefore, the silent time should be synchronized with other COGSIP terminals nearby. The remainder of the frame is a transmission period consisting of a hundred slots. A slot length is 32 μs. A slot consists of 10 μs carrier sensing period and 20 μs data transmission period, and 1 μs guard time is assigned at the end of each period. When the terminal detects a primary system transmission in a 10 μs carrier sensing period, it stops the next transmission of the sub-band that includes the frequency of the detected primary system. Then, the terminal analyzes the detected signal to ascertain whether it is a primary system signal or not because it may be an electromagnetic noise from electronic equipment. The terminal estimates the type of system by a method such as modulation classification (for example, [10]) and if the estimated system is a primary system, the terminal judges that the frequency cannot be used for transmission, and performs a procedure of changing the transmission sub-band. If the estimated system is not a primary system and the power is sufficiently low compared with the COGSIP terminal power, then the terminal judges that the frequency can be used for transmission. In this case, signal transmission in the sub-band is immediately restarted.

3. Key Technologies

To realize the proposed system, a great number of technological problems, including matters concerning implementation, must be solved. In this section, some of these problems and attempted solutions are introduced.

3.1. Carrier Sensing

3.1.1. Sensitivity Improvement Using Overlapped FFT

Carrier sensing is an important research subject because primary system transmission should be protected. When a cognitive terminal is consumer equipment, carrier-sensing methods requiring excessive circuit size and calculation time to obtain high sensitivity cannot be adopted. Although some complicated signal analyzing methods might be adopted, they would be performed at long intervals, and for carrier-sensing that is performed frequently, such as the carrier sensing period before every transmission slot, a simple method should be adopted, for example, energy detection.

At carrier sensing before each slot, the radio wave across the transmission main band is received, A/D converted, divided into FFT (fast Fourier transform) frames and FFTed. By detecting signals by dividing them into FFT bins, many frequencies in the main band are simultaneously detected. In energy detection, multiple FFT frames are defined in a carrier-sensing period, each frame is FFTed and resultant coefficients are transformed into dimension of power. Then, the averaged power of each FFT bin across the multiple frames throughout the carrier-sensing period is calculated. Signal existence is judged by whether the calculated averaged power of each bin exceeds the threshold defined by thermal noise average plus some margin.

Although the energy detection is simple and therefore applicable to wideband signal, the sensitivity is not high. Therefore, we have proposed a high-sensitivity energy detection method to improve both false alarm (FA) probability and misdetection (MD) probability simultaneously when a wideband is A/D-converted at once and Fourier transformed into FFT bins [11].

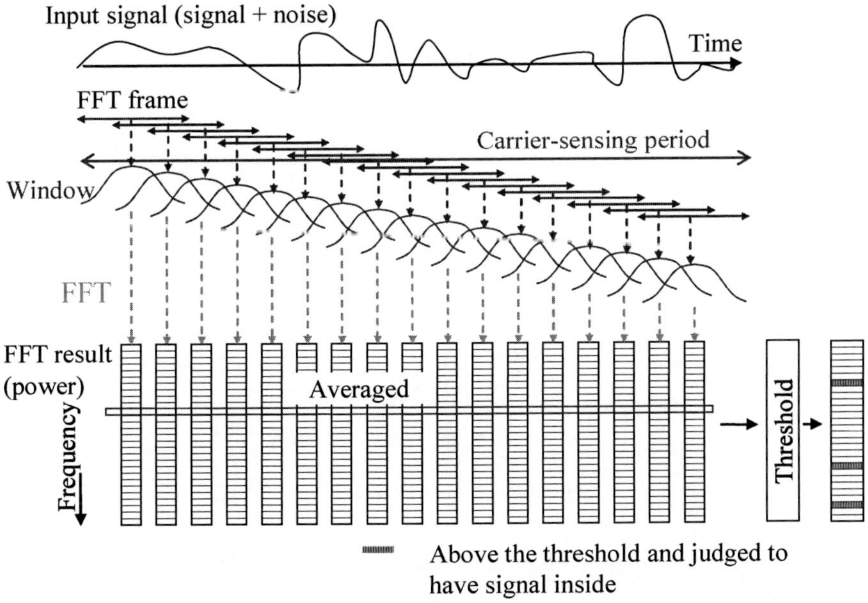

Figure 4. Carrier sensing method using overlapped FFT.

In the proposed method, FFT frames are defined such that adjacent frames overlap greatly with one another as indicated in Figure 4. This technique can be considered a form of sliding window or multi-rate filter bank [12]. The frames are overlapped with larger overlapping rate (the ratio of length overlapping with the adjacent frame to the frame length) than the overlapping rate necessary to keep information and are overlapped with smaller overlapping rate than the full sliding window case. Thus computation amount is much smaller than in the full sliding window case.

In the proposed technique, the averaged power of each FFT bin across all the overlapping frames throughout the carrier-sensing period is calculated and judged against a threshold. As a result, the dispersion of the average due to thermal noise or signal variation can be made smaller than the case of the requisite minimum overlapping.

The necessary overlapping rate to keep information is that with which the minimum composed amplitude of overlapping window functions becomes one. For example, the necessary overlapping rate is 0 for rectangular window. However, with our method, overlapping rate 1/2 can be employed to improve the sensitivity. In our work, it is shown that a noise averaging effect greater than in the case of the necessary overlapping rate can be achieved. Next, the FA and MD probabilities are investigated by computer simulations.

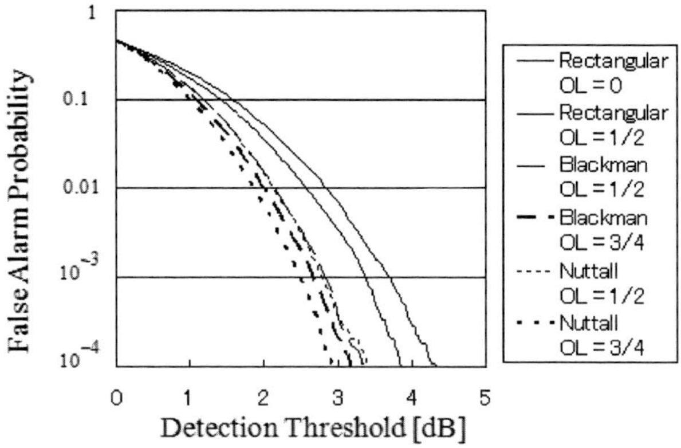

Figure 5. Simulation results of FA probability.

A. False Alarm Probability

In COGSIP, the FFT frame length is 1 μs and actual carrier-sensing length in a carrier-sensing period is nine FFT frame lengths (remaining 1 μs is used for decision). The overlapped FFT frames are defined in this length. For example, when the overlapping rate is 0, the number of frames defined in the period is nine, and when the overlapping rate is 1/2, the number of frames is 17 (=2*8+1).

Figure 5 shows the simulation results of FA probability. AWGN (additive white Gaussian noise) is generated, divided into frames, FFTed, and the resultant power is averaged across the overlapping frames for each bin. The horizontal axis is threshold in dB when the average of thermal noise is 0 dB. The parameters are window shape and overlapping rate (OL). Since the results of the cases OL = 0 do not depend on the window shapes, only the result of rectangular window is indicated.

In carrier sensing of a cognitive radio terminal, many kinds of primary system signals will be received with various timing. When the signal is divided into frames and Fourier transformed, window function with a proper shape with small spectrum spread should be used in order to suppress alias. In Figure 5, three kinds of window functions are used: Blackman, Nuttall and rectangular as reference. The FA probabilities are improved with all the three windows by frame overlapping.

Mathematically, the distribution of power of Fourier-transformed AWGN is exponential. When the power is averaged across the frames of the carrier-sensing period, the distribution becomes gamma distribution $G(x, \sigma^2/p, p)$. x is power (horizontal axis), the scale of distribution is [true value of averaged power of AWGN] / [averaging times] = σ^2/p, and the shape of distribution is averaging times: p. The distribution has a median and is spread around the median. When p is large, the spread becomes small. Using gamma distribution, FA probability is expressed as follows:

$$P_{FA}(x, \sigma^2, p) = 1 - \int_0^x G(z, \frac{\sigma^2}{p}, p)dz \qquad (1).$$

Fitting the values of p of this equation into the results of Figure 5, the effective averaging times can be evaluated corresponding to the overlapping rate and the window shape. The results of fitting are indicated in Table 2. The fitting results in the cases of overlapping rates not shown in Figure 5 are also shown in the table. It can be seen that effective averaging times of over 10 are obtained in 9 frame lengths even with rectangular window. Similarly, for other windows, the effective averaging times keep increasing with the overlapping rate exceeding the necessary overlapping rate. Therefore, it is understood that distribution of noise can be suppressed and the FA probability is improved by overlapping frames greatly with the length of the carrier-sensing period fixed. The effective averaging times are almost saturated at the overlapping rate of from 2/3 to 3/4. The increase of effective averaging times over the saturating value is very small and overlapping rate of 2/3 or 3/4 is sufficient considering the computational load.

Table 2. Effective Averaging Times (Result of fitting the curves of Figure 5 to *p* of equation (1)).

Window shape	Overlapping rate					Necessary overlapping rate
	0	1/2	2/3	3/4	5/6	
Rectangular	9	11.7	12	12	12	0
Blackman	9	17	19.5	20	20	3/5
Nuttall	9	17.5	22	23	23	2/3

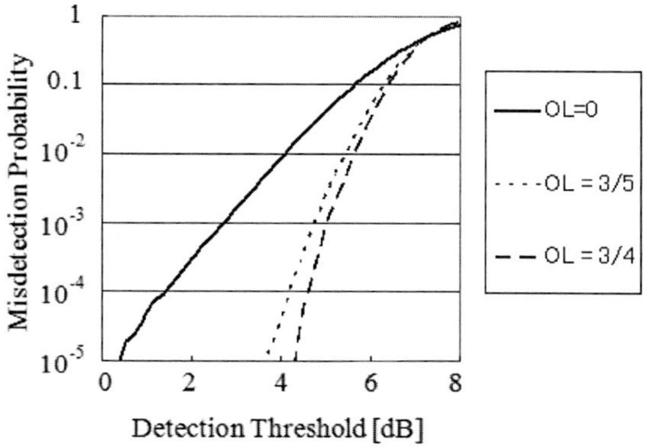

Figure 6. Simulation results of MD probability.

B. Misdetection Probability

Next, a simulation result for MD probability is shown. A QPSK signal was generated in a computer simulation, and MD probability of the signal was evaluated. Two pseudorandom sequences of PN10 were generated and assigned to in-phase (I) and quadrature phase (Q) channels of the QPSK signal, respectively. The generated QPSK signal was eight times oversampled, filtered by root roll-off filter whose rate = 0.5 and distorted slightly by third-order distortion to emulate distortion by RF (radio frequency) system. Then, an AWGN was added. SNR was 6 dB. FFT frames of 32-sample length with Blackman window were defined for the noise-added signal, and each frame was FFTed. FFT bin no. 16, which contains the center frequency part of the generated signal, was carrier sensed by the method indicated in Figure 4. The carrier-sensing period was nine FFT frame lengths. Overlapping rates were 0, 3/5 (necessary overlapping rate for Blackman window) and 3/4.

Figure 6 shows the simulation results for MD probability against detection threshold. The horizontal axis is the detection threshold in dB, the vertical axis is the MD probability and the parameter is the overlapping rate. The detection threshold is relative value when the averaged power of AWGN is set to 0 dB. The MD probability is improved when FFT frames are overlapped. The necessary overlapping rate is 3/5 for Blackman window and the MD probability is better in the case OL = 3/4, which is larger than the necessary overlapping rate, than that in the case of the necessary overlapping rate. Also in the case of MD probability, a similar result to that in the case of FA probability is obtained. The suppression of noise can be achieved in the case of MD probability as well.

This phenomenon and its cause are discussed in detail in [11]. Briefly, it is explained as follows. Since the information in phases of FFT result is dropped at conversion of FFT results into power, when frames of different beginning points are defined, the phase of the same sample contributes differently to the power of each FFT result. Consequently, different power information is produced. From the view of a multi-rate filter bank, a sufficient sampling rate is defined by the bandwidth of center peak of transfer function of the filter and if signal is totally included in the center peak, the defined sampling rate is sufficient. However, the transfer function has a skirt around the peak and when thermal noise or broadband signal is filtered by a narrower bin than the signal itself, part of the signal energy will be exist in the skirt. By increasing the oversampling rate, the signals in the transfer function skirt are averaged. Therefore, the signal is averaged more by overlapping, i.e., oversampling greatly.

3.1.2. Narrow Band Signal Detection Using Autocorrelation

For one of the detection processes during the silent time, narrow-band signal detection using autocorrelation is performed. In COGSIP, the FFT frame length is 1 μs and the effective bandwidth of a bin exceeds 1 MHz. On the other hand, there are many primary signals much narrower than 1 MHz and when energy detection is performed per FFT bin, those narrow-band signals cannot be detected because they do not have sufficient signal-to-noise ratio (SNR) against the bandwidth of an FFT bin. Of course, in many cases, those narrow-band signals have sufficient SNR if it is measured with the bandwidth of narrow-band signal itself.

To detect those narrow-band signals in COGSIP, autocorrelation along with multi-rate filter bank is used [13]. The narrow-band signal detection by autocorrelation can be applied when the energy detection result of an FFT bin has not exceeded the threshold and the primary system assigned to the frequency is known to transmit narrow-band signals.

Figure 7 shows the outline. Similar to the case of Figure 4, overlapped FFT is performed and a series of FFT results for an FFT bin is extracted. In this case, the FFT results are not converted into power but have amplitude and phase. Then the autocorrelation of the extracted samples is calculated. As mentioned above, since FFT can be considered a multirate filter bank, the extracted samples are considered to be the signals filtered by the transfer function of the window function, down-converted to DC and sampled at the rate corresponding to the overlapping rate.

The autocorrelation is calculated as follows [14].

$$R_D(N_s) = \sum_{d=d_0}^{d_0+D-1} \frac{\left| \sum_{n=0}^{N_s-1} x(n)x^*(n-d) \right|^2}{\left(\sum_{n=0}^{N_s-1} |x(n)|^2 \right)^2} \tag{2}$$

n is frame number. *x(n)* is the extracted series of samples. *Ns* is the number of frames to calculate the autocorrelation and *d* is the number of delayed frames. d_0 is the number of minimum delayed frames and it is decided from the shape of the window function and overlapping rate that the noises do not have correlation with the interval. The autocorrelation is normalized by the energy of the whole series and then the judgment threshold does not depend on the power of the signal but only on the autocorrelation property. *D* is the number of averaging varying delayed frame numbers and it leads to less variation of calculated autocorrelation.

A computer simulation was performed with the following condition. A sinusoidal wave of frequency 2048 Hz was generated with 8192 sample/s as a narrow-band signal and AWGN was added. FFT frame length was 16 samples and the window function was Blackman. FFT frame was overlapped with interval of five samples. Since the generated signal was a real number, the effective bin number of FFT result was 8 and the sinusoidal wave existed in the center of the fourth bin. The bin was extracted and autocorrelation was calculated. *Ns* was 256, d_0 was 3 and *D* was 5.

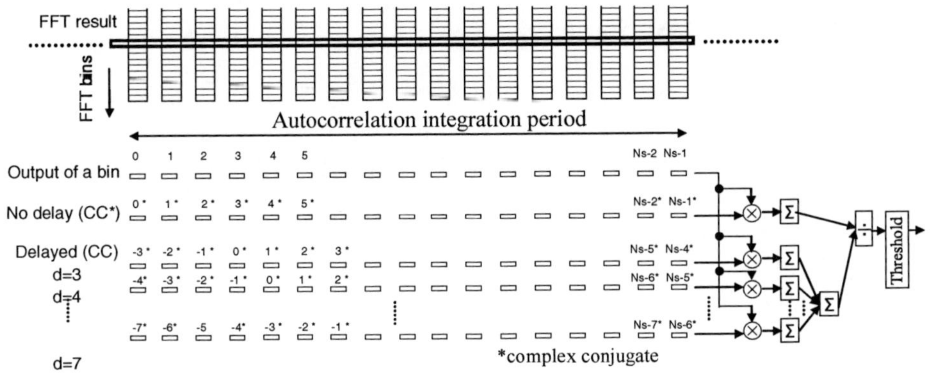

Figure 7. Narrow-band detection using autocorrelation.

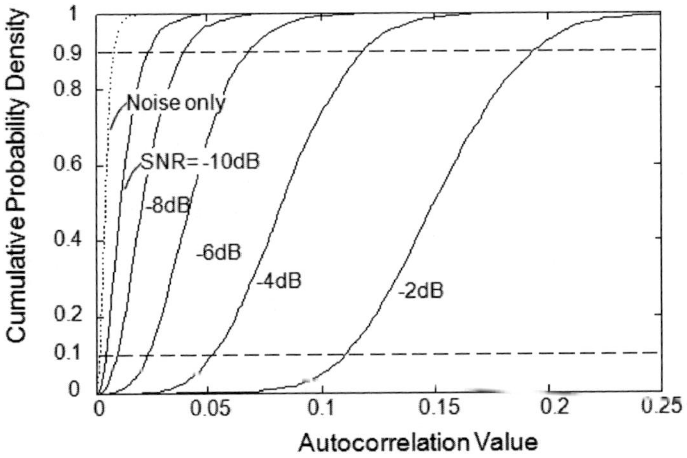

Figure 8. Cumulative probability density of autocorrelation calculation results.

Figure 8 shows the resulting cumulative probability density of calculated autocorrelation values with 2000 trials, varying the SNR of the sinusoidal wave measured with an FFT bin bandwidth. The result shows that if there is SNR of −8 dB, both MD and FA probability of 0.1 can be achieved. For example, when the actual bandwidth of an FFT bin is 3 MHz, narrow-band signal of 50 kHz can be detected if it has SNR of 10 dB with its bandwidth. This method is also applicable to the detection of strong pilot carrier of VSB (vestigial sideband) signal.

3.2. Initial Broadcast Signal

In COGSIP, the frequency of main band is variable among the wide frequency range. It is desirable that the main band frequency and corresponding sub-band frequencies are easily available for a newly joining terminal. Although the easiest way is to prepare a fixed frequency control channel, this method does not seem to be preferable for cognitive systems intended to realize flexible frequency utilization. Therefore, we have proposed a scheme of control channel, actually, initial broadcasting channel, which does not require preliminary knowledge of the frequency of the channel [15]. We call this technique DCPT (differential code parallel transmission).

The process of the technique is indicated in Figure 9. In DCPT, the transmitter transmits a pair of signals whose frequencies are f_1 and f_2 (=f_1+ Δf) that are unknown to the receiver as indicated in Figure 9 (a). The DCPT signal has much narrower bandwidth than that of the main band, and therefore, if it is placed at an arbitrary frequency in the main band, scanning of it will be time-consuming without DCPT detection.

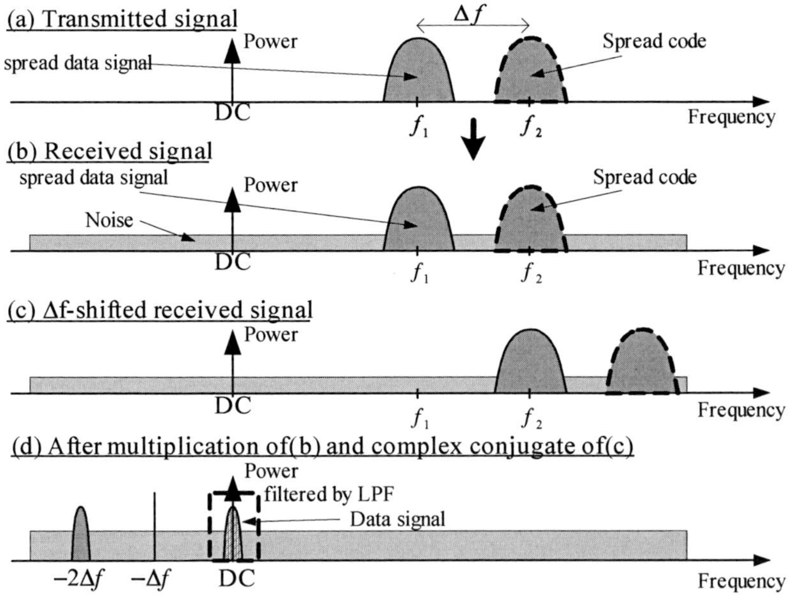

Figure 9. Generation and detection method of the initial broadcast signal.

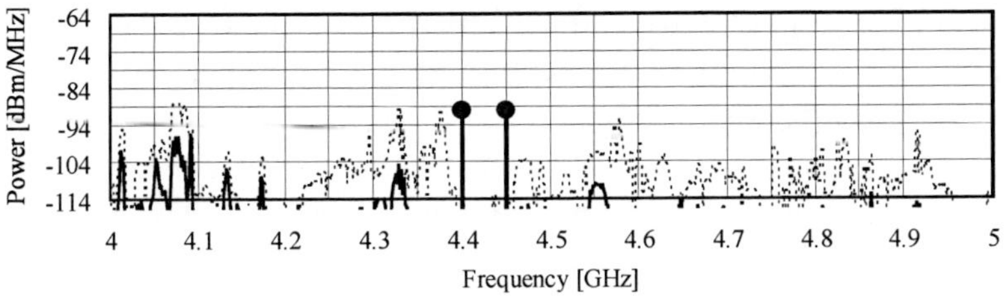

Figure 10. An example of spectrum of interference signals.

One of signals in the pair is a data signal modulated by the broadcasting information and spread using a spreading code. The other is the spreading code itself. The receiver of DCPT signal at first receives the whole main band as indicated in Figure 9 (b). Then, it generates a Δf-shifted received signal as indicated in Figure 9 (c). Next, the receiver multiplies the Δf-shifted signal by the complex conjugate of the signal originally received. Consequently, the despread data signal appears around DC as Figure 9 (d). The signal around DC is filtered by a low-pass filter (LPF) and demodulated. When the receiver knows Δf, it can demodulate the control data signal without the information of f_1 and f_2.

Since it requires a very large spreading factor, the data rate of DCPT channel is very low. Then, the true control channel (which we call 'secondary control channel') exists in another frequency in the main band and, in DCPT channel, only the frequency of the secondary control channel is broadcast. The detailed communications related to the protocol such as joining a network, etc., are carried out in the secondary control channel whose frequency is obtained by the DCPT channel.

Note that in the reception, the frequency of main band needs to be known in advance and only the frequency of detailed DCPT channel whose bandwidth is much narrower than the main band bandwidth can be made unknown. If the receiver does not know the main-band frequency, it must scan the plural main-band frequencies. However, since the process of DCPT reception is very simple, it does not take much time to scan the plural main bands.

We have investigated the performance of DCPT channel by computer simulations. The conditions were as follows. 10 bits of information were transmitted in a frame and a convolutional error correction coding with rate = 1/3 was applied. Tail biting [17] was applied because the bit length after error correction coding was short. Then it was modulated by BPSK and spread 64 times to have the bandwidth of 5 MHz. In order to suppress the influence of multipath fading, the antenna diversity scheme was adopted in which a frame is divided into two parts consisting of the backward and the forward parts and they are transmitted by different antennas. Pilot symbols of two modulation symbols were added in front of each of the two parts. Resulting frame length was 452.2 μs and the data rate was 11.5 kbps.

In DCPT, the receiver receives whole main band and many interference signals exist in the main band. In order to emulate these interference signals, the spectrum indicated in Figure 10 was utilized. We had measured the frequency environment in our office [16] and Figure 10 is the measurement result of the band from 4 to 5 GHz. In the simulation, an AWGN was generated and filtered by the mask of spectrum indicated in Figure 10, and then added to the DCPT signal to emulate interference signals. The ratio of the total interference signal power to the total noise power in the main band was 24.5 dB. The frequency of the DCPT was indicated in Figure 10 by solid and bold lines. The receiver noise figure was 5 dB, and the bandwidth of a main band was 1 GHz.

Simulation results of frame error rate are indicated in Figure 11. In the case without interference signals, the power to achieve 1 % frame error rate corresponds to the SNR of 6.8 dB in the DCPT signal bandwidth, 5 MHz. The degradation by interference signals was only about 1 dB, and the effectiveness of DCPT is confirmed.

In the condition corresponding to Figure 11, the spreading factor was somewhat large in order to achieve sufficient sensitivity suppressing the influence of the noise and interference signals in 1 GHz main band. Therefore, we are investigating the blind adaptive filtering method to lower the spreading factor.

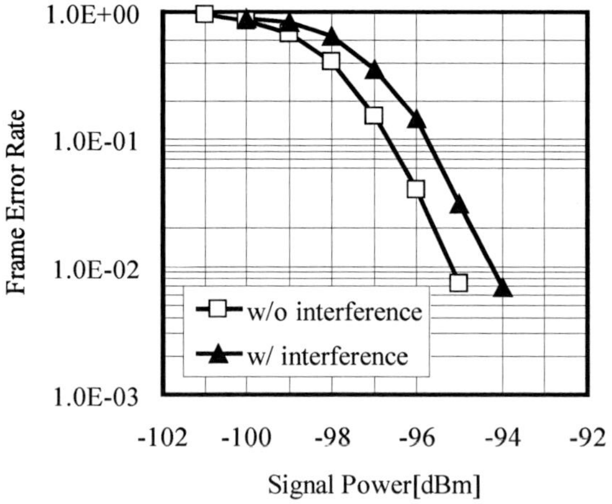

Figure 11. Frame error rate of DCPT signals.

3.3. Transmission Signal Shaping

In COGSIP, many discrete sub-bands in a main band are utilized while avoiding primary transmission. One of the modulation schemes suitable for such usage is OFDM (orthogonal frequency division multiplexing). However, an OFDM signal has high PAPR (peak-to-average power ratio) and the spurious emission may leak out to the frequency of primary transmission due to the distortions in the analogue part. Many approaches have been proposed for suppressing PAPR of OFDM signals. DFT (discrete Fourier transform)-spread OFDM [19] is one of them and is known as a technique that does not require complicated mapping selection nor iteration and the influence on the transmission data itself is small. However, when a DFT-spread OFDM signal is divided into many bands and they are placed in many discrete sub-bands, it does not show the PAPR characteristics as well as in the case it is placed in one band. Therefore, we have proposed a new method that shows better performance in the case of multiple discrete sub-bands [20].

Figure 12 (a) shows the transmitter block diagram of the proposed method. In the method, firstly, the available sub-bands are grouped into K groups. One example of the grouping method is to bundle sub-bands that have similar bandwidth, as indicated in Figure 12 (b). The input bit sequence is divided into K groups corresponding to the total bandwidth of each group. Then, they are modulated by a modulation scheme such as QPSK and converted into frequency region by DFT. After DFT, each symbol sequence is frequency divided to fit the sub-bands included in the group and mapped into subcarriers of an OFDM signal. The mapped subcarriers are then IFFTed and transmitted as conventional OFDM.

A computer simulation was performed. In the simulation, a sub-channel is composed of eight subcarriers and a sub-band was made of from 1 to 20 contiguous sub-channels. In total, 40 sub-channels are randomly allocated in a main band. The bandwidth of a main band was 1024 sub-channels.

Figure 13 shows the simulation result indicating the CCDF (complimentary cumulative distribution function) of PAPR of a signal generated by the proposed method. In order to compare the characteristics, also indicated in Figure 13 are PAPRs of a conventional OFDM signal, a DFT-spread-OFDM signal and a signal in each sub-band of which a single carrier signal is generated. At the point of CCDF 1%, the signal of the proposed method indicates characteristics more than 0.5 dB better than the other signals and it has been found that it indicates good PAPR characteristics.

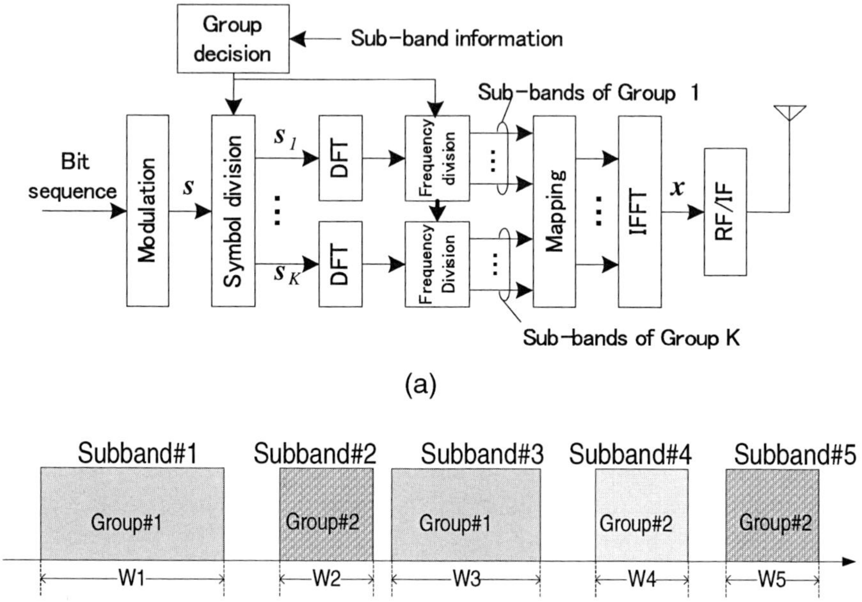

(a) Block diagram of transmitter
(b) Sub-band arrangement example

Figure 12. Transmission signal shaping.

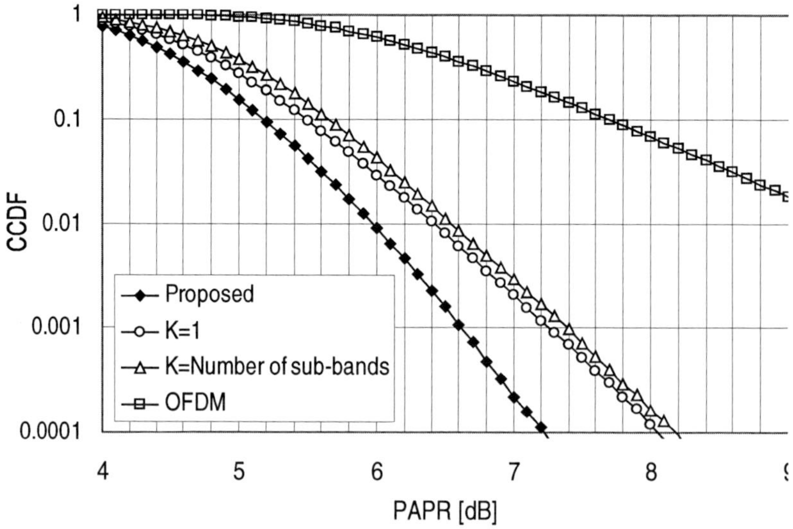

Figure 13. CCDF of PAPR of the signal of the proposed method.

4. Conclusion

The outline of COGSIP that the authors proposed in order to investigate the technical problems of a short-range, high-rate, cognitive system, which utilizes multiple white spaces simultaneously, is introduced and some of the investigated techniques are introduced. COGSIP is a future system and there are many problems to be solved. In this chapter, we have introduced a high-sensitivity energy detection method, a narrow-band primary signal detection method, an initial broadcast signal that can be easily detected without knowledge of the frequency, and a transmission signal generation method with reduced PAPR.

References

[1] Mitola III, J. (1999). "Cognitive Radio for Flexible Mobile Multimedia Communications, " *MoMuC* **'99**, 3-10.

[2] Haykin, S. (2005). "Cognitive Radio: Brain-Empowered Wireless Communications," *IEEE J. Select. Areas Commun.*, Vol.23, no.2, 201-220, Feb. 2005.

[3] Cabric, D., O'Donnell, I. D., Chen, M. S. W. & Brodersen, R. W. (2006). "Spectrum Sharing Radios," *IEEE Circuits And Systems Magazine*, 2nd Quarter, 30-45.

[4] http://hraunfoss.fcc.gov/edocs_public/attachmatch/DOC-286566A1.pdf

[5] Ofcom, "Digital Dividend Review", http://www.ofcom.org.uk/consult/condocs/ddr/statement/

[6] http://grouper.ieee.org/groups/802/22/

[7] https://mentor.ieee.org/802-sg-whitespace/documents

[8] http://gullfoss2.fcc.gov/prod/ecfs/retrieve.cgi?native_or_pdf=pdf&id_document=6519868157

[9] Tomioka, T., Horiguchi, T., Sakata, R., Tomizawa, T. & Inoue, K. (2007). "A Proposal of High-Flexibility Short-Range Cognitive Radio System", *International Symposium on Signals, Systems and Electronics*, 95-98.

[10] Nandi, A. K. & Azzouz, E. E. (1998). "Algorithms for automatic modulation recognition of communication signals," *IEEE Trans. Comm.*, Vol. 46, No. 4, 431-436.

[11] Tomioka, T., Tomizawa, T. & Kobayashi, T. (2009). "High-Sensitivity Carrier Sensing Using Overlapped FFT for Cognitive Radio Transceivers," *2009 IEEE 69th Vehicular Technology Conference*, 10P-4 April 2009.

[12] Malvar, H. S. (1992). "Signal Processing with Lapped Transforms," Chapter 3, Artech House.

[13] Tomioka, T., Sakata, R. & Kobayashi, T. (2008). "A Narrow-Band Signal Detection Technique Using Autocorrelation for a Cognitive Radio Terminal," (In Japanese) IEICE Technical Report, SR2008-5.

[14] Polydoros, A. & Woo, K. T. (1985). "LPI detection of frequency-hopping signals using autocorrelation techniques," IEEE JSAC, Vol. Sac-3, No.5, 714-726.

[15] Sakata, R., Tomioka, T. & Kobayashi, T. "A Cognitive Radio Transmission Scheme without the Carrier Frequency Information," IEICE Technical Report, SR2007-92 (In Japanese)

[16] Horiguchi, T., Sakata, R., Inoue, K., Tomizawa, T. & Tomioka, T. "Measurement Results of Spectrum Usage for Cognitive Radio System" IEICE Technical Report, SR2006-86 (In Japanese)

[17] Ma, H. & Wolf, J. (1986). "On tail biting convolutional codes," *IEEE Trans. Commun.*, Vol. 34, Issue 2, 104-111.

[18] Sakata, R., Tomioka, T. & Kobayashi, T. (2009). "An Improvement of Broadcast Signal Transmission Scheme for Cognitive Radio System," *General Conference of IEICE*, B-17-10 (In Japanese).

[19] Galda, D. & Rohling, H. (2002). "A low complexity transmitter structure for OFDM-FDMA uplink systems," *VTC Spring 2002*, Vol. 4, 1737-1741.

[20] Ban, K., Tomioka, T. & Kobayashi, T. (2008). "Transmitter Design for Cognitive Radio Systems under Distributed Frequency Subbands Assignment"、General Conference of IECIC, B-8-3 (In Japanese).

In: Cognitive Radio
Editors: P.D. McGuire et al, pp. 121-138
ISBN: 978-1-60876-604-8

Chapter 5

SPECTRUM ADAPTATION BASED MIMO-OFDM SYSTEM FOR DYNAMIC SPECTRUM SHARING

Hao Hu [*] ***and Junde Song***
Dept. of Electronic Engineering, Beijing University of Posts and Telecommunications, Beijing, China, 100876

Abstract

In this paper, a novel dynamic spectrum sharing scheme combining spectrum adaptation and MIMO-OFDM for cognitive radio system is proposed. Given spectrum sensing and channel estimation information by the receiver, an improved model due to transmit power thresholds for SA-MIMO-OFDM system is adopted to achieve spectrum adaptation for unlicensed users. In order to improve spectrum efficiency and reduce computational complexity, a simple power allocation algorithm is also proposed. Simulations indicate that the scheme solves the partial interference problem in Interference Temperature Model. The BER performance of the 4Tx4Rx SA-MIMO-OFDM system is over 7 dB higher than the SISO system without spectrum sharing. Moreover, power threshold reduction has little impact on the BER performance of the proposed system with different antenna numbers. Compared to SISO system using Power Threshold Model, the spectral efficiency of the proposed system with 2Tx2Rx MIMO increases 7.8 bps/Hz at a target SNR=30 dB, and the loss in spectral efficiency

---

[*] E-mail address: cnhuhao@gmail.com. (Corresponding author)

with different power threshold reduction is at most 1.5 bps/Hz at average SNR of 30 dB.

Keywords: Dynamic spectrum sharing; MIMO-OFDM; cognitive radio; spectrum adaptation; dynamic power allocation.

I. Introduction

As access to available spectrum is becoming increasingly difficult, many techniques about improving the spectrum utilization have been presented. Cognitive radio[1][2], defined as an intelligent wireless communication system that is aware of its environment and adapts its statistical variations according to learning from it, is viewed as an efficient approach for improving the utilization of the precious spectrum resource. Recent researchers have presented two basic dynamic spectrum sharing approaches in cognitive radio. One, based on hard constraints, prohibits unlicensed users from accessing the licensed frequency band, and the other based on soft, constrains unlicensed signal power below a predefined limit, such as the Interference Temperature Model (ITM) proposed by the FCC [3][4]. However, the partial interference problem of ITM may introduce severely interference to licensed users.

Orthogonal frequency division multiplexing (OFDM) is a promising candidate for cognitive radio system to provide high data rates while avoiding interference with licensed user transmission. Meanwhile, multiple-input and multiple-output (MIMO)[5] is a powerful technology with significant data rate improvements over single-input single-output (SISO) system without wasting spectrum resource. In this paper, a variant of MIMO-OFDM called spectrum-adaptation OFDM (SA-MIMO-OFDM) is proposed, where the implementation achieves high data rates via a novel dynamic power allocation algorithm based on Power Threshold Model, solving the partial interference problem of Interference Temperature Model [3].

The rest of the paper is organized as follows: Section II introduces the system framework, channel model and detection theme of SA-MIMO-OFDM system. Section III provides the mathematical model of the novel spectrum sharing scheme. Section IV formulates the problem of data rate maximization under power thresholds and presents the proposed algorithm. Simulation results are presented in Section V, and several concluding remarks are made in Section VI.

II. System Description

A. System Framework

The block diagram of SA-MIMO-OFDM system, which has N_{tx} transmit antennas and N_{rx} receive antennas, is shown in Figure 1. A high speed input data stream is modulated using *M*-ary quadrature amplitude modulation (*M*-QAM). After MIMO precoding, the modulated data stream is split into N_{tx} data streams by a serial-to-parallel (S/P) converter. After applying inverse fast Fourier transform (IFFT) to the N_{tx} data streams, a guard interval with a length greater than the channel delay spread is added in order to mitigate the effect of intersymbol interference (ISI). The baseband SA-MIMO-OFDM signal is transmitted to the radio environment via N_{tx} antennas.

At the receiver side, the received signals from all the antennas will then pass through a Fast Fourier Transform (FFT) with the removal of the cyclic prefix. Then, the received signal will pass through the MIMO decoder and M-QAM Demodulator to retrieve the original transmitted signal.

B. Channel Model

We consider a MIMO-OFDM system with N_{tx} transmit antennas and N_{rx} receive antennas. Let *N* be the total number of subcarriers. The MIMO channel model for the *k*th subcarrier can be represented by

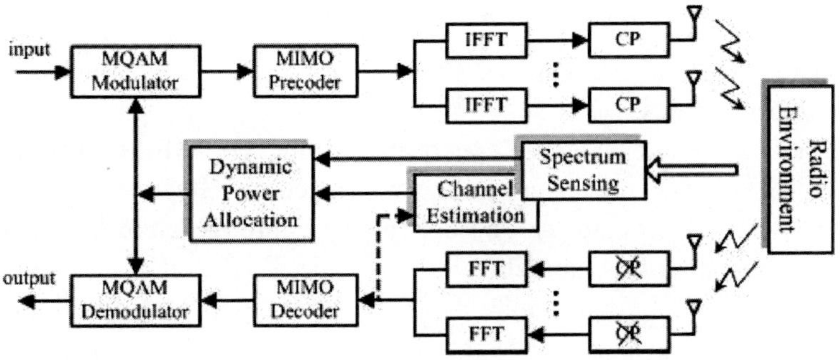

Figure 1. System Framework of SA-MIMO-OFDM.

$$\mathbf{Y}_k = \mathbf{H}_k \mathbf{X}_k + \mathbf{n}_k \qquad k = 1, 2, \cdots, N \tag{1}$$

where $\mathbf{X}_k = \left[x_{k,1}\ x_{k,2} \cdots x_{k,tx} \right]^T$ and $\mathbf{H}_k \in C^{N_{rx} \times N_{tx}}$ are transmitted signal vector and discrete-time channel response of the kth subcarrier, respectively. $\mathbf{n}_k$ is an AWGN vector with covariance matrix $\sigma_n^2 \mathbf{I}_{N_{rx}}$.

Using the singular value decomposition (SVD) [6] the channel matrix on the kth subcarrier can be written as $\mathbf{H}_k = \mathbf{U}_k \mathbf{\Lambda}_k \mathbf{V}_k^H$, where $\mathbf{U}_k \in C^{N_{rx} \times N_{rx}}$ and $\mathbf{V}_k^H \in C^{N_{tx} \times N_{tx}}$ are unitary matrices, and $\mathbf{\Lambda}_k \in C^{N_{rx} \times N_{tx}}$ is a diagonal matrix consisting of singular values of $\mathbf{H}_k$ which is defined as

$$\begin{gathered} \mathbf{\Lambda}_k = diag(h_{k,1}, h_{k,2}, \cdots, h_{k,r}, 0, \cdots, 0) \\ h_{k,1} \geq h_{k,2} \geq \cdots \geq h_{k,r} \geq 0 \end{gathered} \tag{2}$$

If the channel state information at the transmitter and receiver is perfect and joint optimal precoder and decoder satisfy $\mathbf{R}_k = \mathbf{U}_k$, $\mathbf{T}_k = \mathbf{V}_k$, Eq. (1) can be rewritten as

$$\mathbf{Y}_k = \mathbf{R}_k^H \mathbf{H}_k \mathbf{T}_k \mathbf{X}_k + \mathbf{R}_k^H \mathbf{n}_k = \mathbf{\Lambda}_k \mathbf{X}_k + \mathbf{n}'_k \tag{3}$$

where $\mathbf{n}'_k = \mathbf{R}_k^H \mathbf{n}_k$ and the distribution of $\mathbf{n}'_k$ is invariant under unitary transformation.

Thus, we have decomposed the MIMO channel for a particular subcarrier into $R \leq \min(N_{tx}, N_{rx})$ parallel sub-channels, and $h_{k,r} (r = 1, 2, \cdots, R)$ are channel gains of the equivalent subchannels. There are totally $M = RN$ sub-channels in the MIMO-OFDM systems.

In order to achieve dynamic spectrum sharing, we will set different power thresholds for all the subcarriers in the next section. Given the decomposition outlined above, this problem can be decomposed into dynamic power allocation over all the equivalent subchannels.

C. Detection Theme

In the proposed system, we use the minimum mean square error (MMSE) linear detection to obtain better performance. The MMSE detector minimizes the mean square error between the actually transmitted symbols and the output of the linear detector which is defined by

$$\mathbf{G}_{MMSE} = \mathbf{H}^H \left(\sigma^2 \mathbf{I}_{N_r} + \mathbf{H}^H \mathbf{H}\right)^{-1} \tag{4}$$

where $\mathbf{G}$ denotes the matrix representation of the detection scheme. The estimation error of the different layers corresponds to the main diagonal elements of the error-covariance matrix, which can be defined as

$$\mathbf{\Phi} = \sigma^2 \left(\mathbf{H}^H \mathbf{H} + \sigma^2 \mathbf{I}_{N_r}\right)^{-1} \tag{5}$$

III. Dynamic Spectrum Sharing Model

The interference temperature model, which was proposed for dynamic spectrum management [4] by FCC, is used for measuring the power and bandwidth occupied by interference. The interference temperature limit T_L is a maximum amount of tolerable interference for a given frequency band in a particular location. Any unlicensed transmitter utilizing licensed band must guarantee that their transmissions added to the existing interference must not exceed the interference temperature limit at a licensed receiver.

$$P_T(f_c) = \delta P_S(f_c) + P_I(f_c) \leq BkT_L(f_c) \tag{6}$$

where $P_S(f_c), P_I(f_c), T_L(f_c)$ is the unlicensed signal transmit power, interference power and interference temperature limit, which are centered at f_c, covering bandwidth B. δ is the propagation loss factor between unlicensed transmitter and licensed receiver, $\delta \in (0,1)$. δ can be measured via spectrum sensing techniques, which is not involved in this paper.

However, the Interference Temperature Model does not regulate absolute interference but average interference. Thus, the absolute interference would exceed the interference temperature limit in some frequency bands, while the average interference over the licensed signal may be under the threshold. This partial interference problem will introduce severely interference to licensed users.

In this section, an improved model based on power threshold (PTM) is proposed to solve this problem, which has a stronger requirement.

$$P_S(f) \leq \left[BkT_L(f_c) - P_I(f)\right] / \delta \tag{7}$$

Note that this requirement implies that the total interference at the licensed receiver is under the interference temperature limit.

$$P_T(f_c) \leq \max\left[\delta P_S(f) + P_I(f)\right] \leq BkT_L(f_c) \tag{8}$$

According to (8), we can solve the partial interference problem via changing the unlicensed transmit power.

In the OFDM system, we assume that ΔB is the bandwidth of each subcarrier, and $f_{ini} = f_c - B/2$, the average transmit power threshold of unlicensed signal for kth subcarrier is defined as

$$P_{TH}(k) = \left[BkT_L(f_k) - \frac{N}{B}\int_{f_{ini}+k\Delta B}^{f_{ini}+(k+1)\Delta B} P_I(f)\,df\right] \Big/ \delta \tag{9}$$

where $B_{lic} \subset B$ is the frequency band where licensed signals are detected , $T_L(f_k)$ is a function of frequency measured on the kth subcarrier, which is defined by

$$T_L(f_k) = \begin{cases} T_{\min} \propto \dfrac{kB}{P_I(f_k) + \delta P_S(f_k)} & f_k \in B_{lic} \\ T_{\max} \propto \dfrac{kB}{P_I(f_k)} & f_k \notin B_{lic} \end{cases} \tag{10}$$

where T_L is chosen to be maximum or minimum according to the existence of the licensed signal, which can achieve spectrum adoption and increase the data rate of the unlicensed signal without interference.

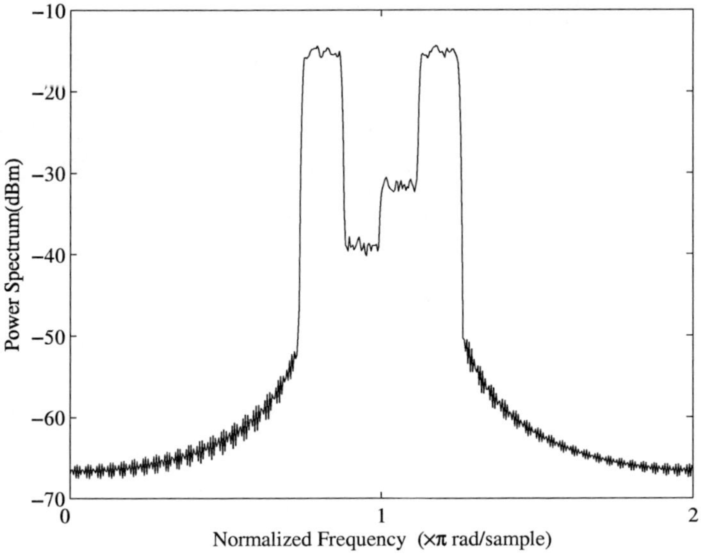

Figure 2. Power spectrum waveform of OFDM with spectrum adaptation.

In order to maximize both capacity and spectral efficiency while minimizing absolute interference, the unlicensed signal transmit power on each subchannel should be equal to the power threshold.

$$P_S(f_k) = P_{TH}(k) \tag{11}$$

According to (9) and (11), we can set power thresholds over each subcarrier to achieve spectrum adaptation. Figure 2 illustrates the power spectrum of OFDM with spectrum adaptation based on Power Threshold Model.

IV. Dynamic Power Allocation Algorithm

In this section, we propose a simple and efficient dynamic power allocation algorithm that maximizes data rate of the proposed MIMO-OFDM system under the constraints of total transmit power and average power threshold in each subcarrier.

A. Problem Formulation

According to the analysis in Section , N subcarriers of the OFDM system can be decomposed into $M=RN$ equivalent subchannels. Assumed that the target average total power is P_{budget} and maximum tolerable BER is $BER_{\max}$, the power gain and the variance of the additive white Gaussian noise of the rth subchannel of the kth subcarrier as $h_{k,r}$ and $\sigma_{k,r}^2$, respectively.

If the transmit power allocated to the rth subchannel of the kth subcarrier is $P_{k,r}$, the corresponding number of loaded bits can be obtained by

$$b_{k,r} = \log_2\left(1+\frac{P_{k,r}\cdot h_{k,r}{}^2}{\sigma_{k,r}^2\Gamma}\right) \tag{12}$$

where Γ is SNR gap that is calculated according to the "gap approximation" analysis [7] based on the modulation scheme used (M-QAM in this case), the desired BER and the system performance margin, which can be written as

$$\Gamma = -\frac{\ln(5\cdot BER)}{1.5} \tag{13}$$

Then the problem considered in this paper can be formulated as to maximum the transmit bit numbers under the constraints of P_{budget}, P_{TH} and $BER_{\max}$.

$$\begin{aligned} &\text{maximize} \quad B=\sum_{k=1}^{N}\sum_{r=1}^{R} b_{k,r} \\ &\text{subject to} \quad \sum_{k=1}^{N}\sum_{r=1}^{R} P_{k,r} \le P_{budget} \\ &\qquad\qquad \sum_{r=1}^{R} P_{k,r} \le P_{TH}(k) \quad \forall k\in\{1,2,\cdots,N\} \\ &\qquad\qquad BER_k \le BER_{\max} \quad \forall k\in\{1,2,\cdots,N\} \end{aligned} \tag{14}$$

where $P_{TH}(k)$ is the power threshold of the kth subcarrier.

The problem above is still a convex problem since the additional power constraint is convex, thus the solution using a Lagrange multiplier technique will be optimal.

$$F(\boldsymbol{P},\lambda_1,\lambda_2)=\sum_{k=1}^{N}\sum_{r=1}^{R}b_{k,r}+\lambda_1\left(\sum_{k=1}^{N}\sum_{r=1}^{R}P_{k,r}-P_{budget}\right)+\lambda_2\left(\sum_{r=1}^{R}P_{k,r}-P_{TH}(k)\right) \quad (15)$$

where λ_1, λ_2 are Lagrange multiplier factors. Then the problem can be solved by the greedy algorithm with "gap approximation" [8]. However, the greedy algorithm heavily burdens the computational load because of its extensive sorting operations in all the subchannels. In addition, how to allocate transmit power under constraints efficiently is still a problem.

B. Proposed Algorithm

In this subsection, we present a simple dynamic power allocation algorithm with "gap approximation" under the constraints of transmit power. Firstly, the same transmit power is assigned to each subchannel, that is

$$P_{ini}=P_{budget}/M=P_{budget}/(N\cdot R) \quad (16)$$

According to (12), the bit allocation matrix $\boldsymbol{b}_{ini}\in C^{N\times R}$ is calculated and we round off to make $b_{k,r}$ be an integer number, which can be expressed as

$$b_{k,r}=round\left[\min\{b_{ini}(k,r),\gamma\}/d\right]\times d \quad (17)$$

where γ is the largest modulation level of the proposed system and d is the step length of bit allocation according to the modulation scheme. Then, the initial transmit power allocation matrix is calculated according to (12) and (17).

In order to reduce the total power of N subcarriers under the power constrains, the bit allocation must be adjusted over each subcarrier. The adjustment of the bit allocation for R subchannels of the kth subcarrier through following three steps:

Step 1 Calculate the power reduction and increment when change d bits allocation for each subchannel.

$$\Delta P^-(b_{k,r}) = P(b_{k,r}) - P(b_{k,r} - d) \tag{18a}$$

$$\Delta P^+(b_{k,r}) = P(b_{k,r} + d) - P(b_{k,r}) \tag{18b}$$

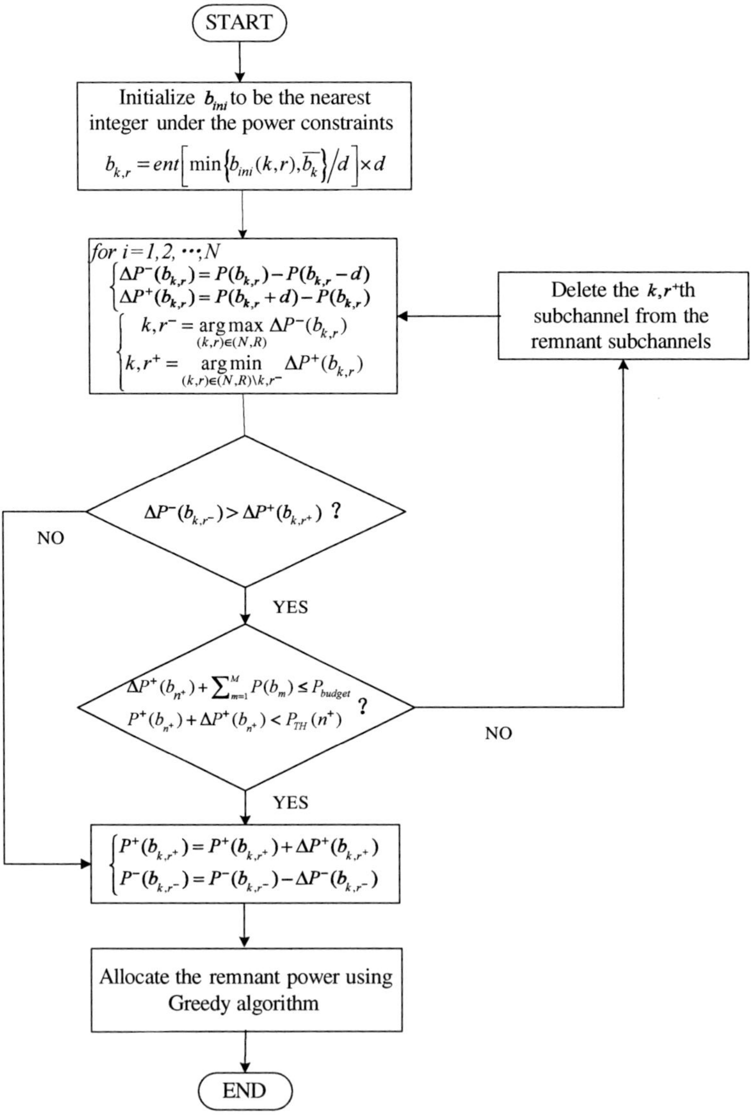

Figure 3. Flow chart of the proposed algorithm.

Step 2 Exclude the subchannel with the maximal power reduction, and find out the subchannel with the minimal power augment in the remainders.

$$\begin{aligned} k,r^{-} &= \underset{(k,r)\in(N,R)}{\arg\max}\ \Delta P^{-}(b_{k,r}) \\ k,r^{+} &= \underset{(k,r)\in(N,R)\backslash k,r^{-}}{\arg\min}\ \Delta P^{+}(b_{k,r}) \end{aligned} \tag{19}$$

Step 3 If $\Delta P^{-}(b_{k,r^{-}}) < \Delta P^{+}(b_{k,r^{+}})$, then STOP. If the power threshold condition of *k*th subchannel and the total transmit power constraint are satisfied, the transmit power adjustment is accomplished as follows:

$$P^{+}(b_{k,r^{+}}) = P^{+}(b_{k,r^{+}}) + \Delta P^{+}(b_{k,r^{+}}) \tag{20a}$$

$$P^{-}(b_{k,r^{-}}) = P^{-}(b_{k,r^{-}}) - \Delta P^{-}(b_{k,r^{-}}) \tag{20b}$$

Otherwise, we exclude the r^{+}th subchannel from the remnant subchannels and return to **Step 1**.

After above steps completed over all the subcarriers, the remnant transmit power can be allocated by the Greedy algorithm under the total power constraint and transmit power thresholds. The flow chart of the above algorithm is shown as follows.

V. Simulation Results

In this section, simulation results are presented to compare the performance of the various spectrum sharing approaches introduced in this paper. For the simulations, N=256 MQAM-modulated (M=4, 16, 64, 256) subcarriers are employed and the system sampling interval is 50 ns and the length of cyclic prefix is 32. To compare those results properly, we set the period per OFDM symbol to be 14.4 us and the interval of subcarriers to be 78.125 kHz. In this paper, it is assumed that channel estimation and spectrum sensing is perfect and the delay of the feedback of channel state information to the transmitter is negligible. The detailed simulation parameters are given in Table 1.1

Table I. System Parameters for SA-MIMO-OFDM Platform.

Sampling interval	0.05*us*
Modulation Type	0, 4-, 16-, 64-, 256-QAM
Total no. of subchannels (N_C)	256
IFFT/FFT length	256
Cyclic prefix length	64
OFDM symbol period	16 μs
Subchannel interval	78.125kHz
Total Bandwidth	20MHz
Channel coding	None
No. of transmit antenna N_T	2, 4
No. of receive antenna N_R	2, 4

In the simulations, European wireless local network standard HIPERLAN/2's channel model C[12] which has a multi-path (14-path) delay profile is considered. The delay profile of channel model C is shown in Table II and Figure 4 depicts the Instantaneous state characteristic of HIPERLAN/2 Channel Model C.

Table II. Delay Profile of HIPERLAN/2 Channel Model C.

Path	Delay (ns)	Power (dB)	Doppler Profile
1	0	-3.3	Clarke
2	10	-3.6	Clarke
3	20	-3.9	Clarke
4	30	-4.2	Clarke
5	50	-0	Clarke
6	80	-0.9	Clarke
7	110	-1.7	Clarke
8	140	-2.6	Clarke
9	180	-1.5	Clarke
10	230	-3	Clarke
11	280	-4.4	Clarke
12	330	-5.9	Clarke
13	400	-5.3	Clarke
14	490	-7.9	Clarke

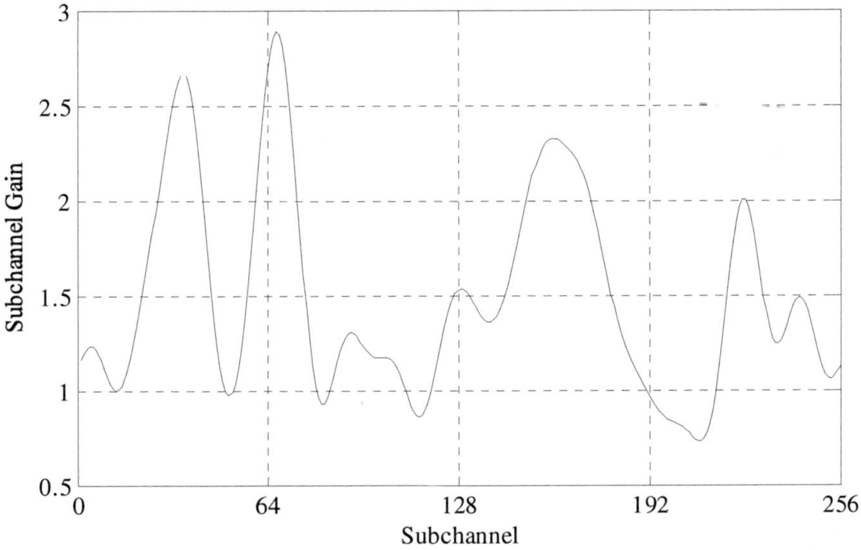

Figure 4. Instantaneous state characteristic of HIPERLAN/2 Channel Model C.

Assumed that the licensed frequency band $B = [0, 20MHz]$ and the licensed signal exits on $B_{lic} = [5, 15MHz]$. Let N be the total number of subcarriers, and N_{lic} is the total number of subcarriers on B_{lic}. For a given frequency band in a particular location, the power thresholds for all the subcarriers are calculated according to spectrum detection and analysis. In the Power Threshold Model, we define the measurement of average power constraint on B_{lic} by power threshold factor η, which can be expressed as follows:

$$\eta = 1 - \frac{N \sum_{f_k \in B_{lic}} P_{TH}(k)}{N_{lic} P_{budget}} \tag{21}$$

For the simulations, none spectrum sharing scheme and spectrum sharing based on ITM and PTM with variable power threshold reduction factor (η = 0%, 20% and 40%) are used to compare the system performance.

Figure 5 shows the power allocation of two spectrum sharing schemes with different algorithms. From these two curves, we can find that the peak power of the unlicensed signal exceeds the interference temperature limit on some subcarriers, using Greedy algorithm [8] based on ITM. However, it can be observed that our algorithm can generate shaped power spectrum via proper power thresholds, which results the same allocation to the optimal Greedy algorithm in the unoccupied licensed band with maximum capacity.

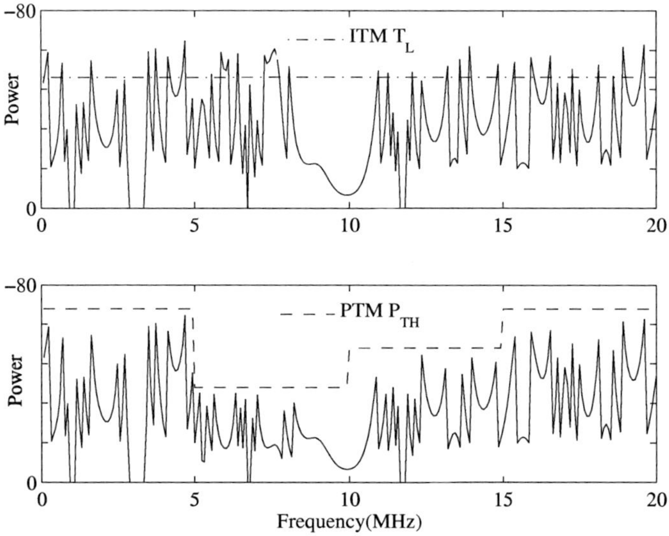

Figure 5. Power Allocation; spectrum sharing with ITM and PTM.

Figure 6 depicts bit error rate (BER) performance for none spectrum sharing scheme (None) and spectrum sharing based on ITM and PTM, respectively. The BER performance for PTM with single-antenna and multi-antennas are also shown for comparison. It is clear that although Figure 6 shows the spectral efficiency of PTM is a little lower than ITM scheme in SISO system, the BER performance increases by about 3 dB as a result of partial interference in ITM scheme. Using MIMO technique with 4 transmit antennas and 4 receive antennas, the BER performance of PTM scheme increases over 7.5 dB. Compared to none spectrum sharing theme, the improvement of the BER performance of 4Tx4Rx SA-MIMO-OFDM system is over 7 dB.

The BER performance of the PTM using MIMO with variable power threshold factors is shown in Figure 7. The power threshold factor η is assumed to be 0% (None), 20% and 40%, respectively. Three themes are simulated in the experiments, such as SISO, MIMO with 2Tx2Rx and MIMO with 4Tx4Rx. We observe that reducing the power thresholds has little impact on the BER performance of the proposed system with different antenna numbers.

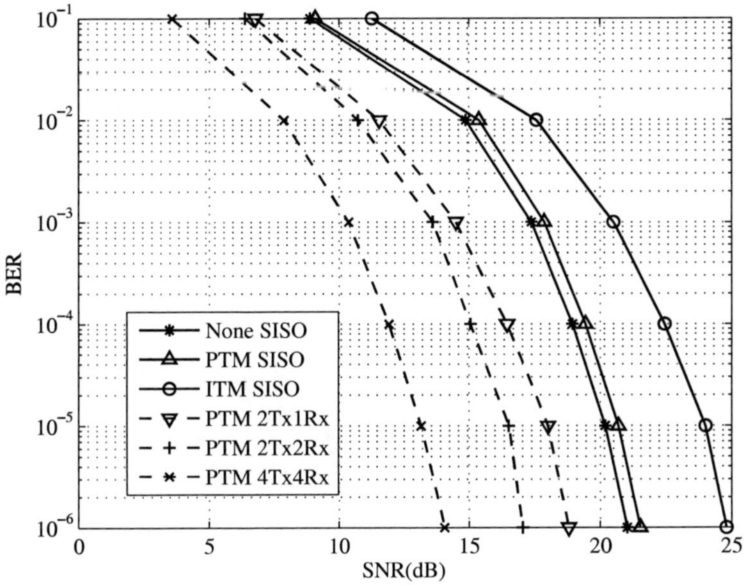

Figure 6. BER versus SNR; none spectrum sharing, spectrum sharing with ITM and PTM.

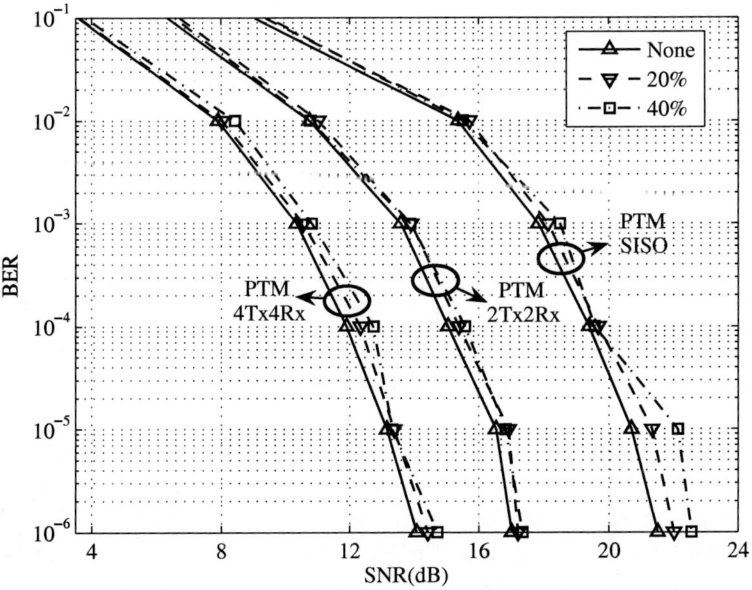

Figure 7. BER versus SNR; PTM with variable power threshold factors.

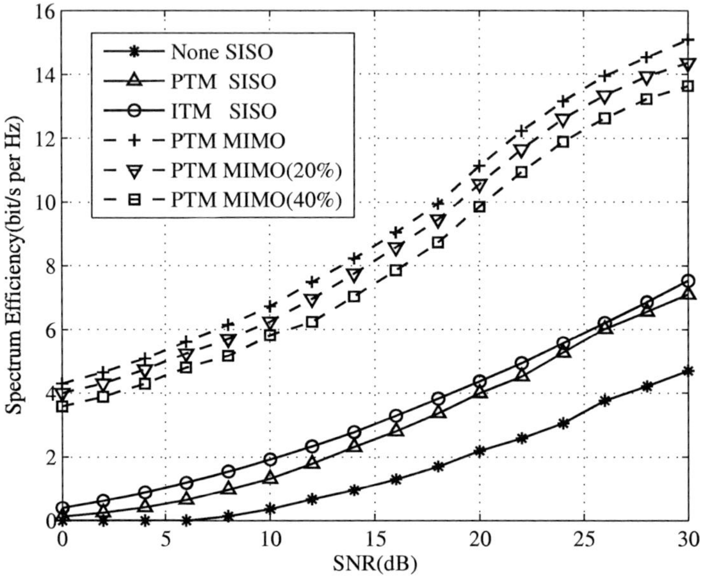

Figure 8. Spectrum Efficiency; none spectrum sharing, spectrum sharing with ITM and PTM (BER=10^{-3}).

In Figure 8, we evaluate the spectrum efficiency of three scenarios, which are none spectrum sharing scheme (None), spectrum sharing schemes based on ITM and PTM. The PTM with different antenna numbers are compared under variable power threshold constraints. It is shown that the spectral efficiency of PTM is a little lower than ITM scheme due to more strict power constraint in SISO system at a target BER=10^{-3}. However, the PTM with MIMO (2Tx2Rx) can greatly improve the spectral efficiency of 7.8 bps/Hz at a target SNR=30 dB. When the power threshold factor η is changed into 20% and 40%, there is at most 1.5 bps/Hz loss in spectral efficiency at average SNR of 30 dB.

VI. Conclusion

In this paper, we present a dynamic spectrum sharing scheme based on SA-MIMO-OFDM for cognitive radio systems. In order to use the unlicensed signal power efficiently, a dynamic power allocation algorithm is also proposed. Simulation results show that the propositional scheme solves the partial interference problem of Interference Temperature Model and improves the spectrum utilization. The BER performance of the 4Tx4Rx SA-MIMO-OFDM system is over 7 dB higher than the SISO system without spectrum sharing. Moreover, power threshold reduction has little impact on the BER performance of the proposed system with different antenna numbers. Also, compared to SISO system using Power Threshold Model, the spectral efficiency of the proposed system with 2TxRx increases 7.8 bps/Hz at a target SNR=30 dB, and the loss in spectral efficiency with different power threshold reduction is at most 1.5 bps/Hz at average SNR of 30 dB. All the results indicate that this proposed system can achieves dynamic spectrum sharing and performs better than the traditional methods due to its higher spectrum efficiency.

Acknowledgment

This work was supported by the Science-Technology Supporting Project of the National "Eleventh Five-Year-Plan" of China under Grant No. 2006BAH02A01.

References

[1] Mitola III, J. & Maguire, G. Q. (1999). Cognitive radio: making software radios more personal. *IEEE Pers. Commun.*, vol. 6, no. 4, 13-18.

[2] Haykin., S. (2005). Cognitive Radio: Brain-Empowered Wireless Communications. *IEEE Journal on Selected Area in Communications*, vol.23, no 2, 201-220, Feb. 2005.

[3] Clancy, T. & Walker, B. (2006). Spectrum shaping for interference management in cognitive radio networks. *SDR Forum Technical Conference 2006.*

[4] FCC. Establishment of Interference Temperature Metric to Quantify and Manage Interference and to Expand Available Unlicensed Operation in Certain Fixed Mobile and Satellite Frequency Bands. ET Docket 03-289, Notice of Inquiry and Proposed Rulemaking, 2003.

[5] Stuber, G. L., Barry, J. R., Mclaughlin, S. W., et al. (2004). Broadband MIMO-OFDM wireless communications. *Proceedings of the IEEE*, vol. 92, no. 2, 271-294, Feb 2004.

[6] Gao, J. & Faulkner, M. (2002). On implementation of bit-loading algorithms for OFDM systems with multiple-input multipleoutput. *Proc. of IEEE Veh. Technol. Conf. (VTC) '02-Fall*, 199-203, Sept 2002.

[7] Cioffi, J. M., et al. (1995). MMSE Decision-Feedback Equalizers and Coding Part I: Equalization Results. *IEEE Trans. Commun.*, vol. 43, no 10, 2582-2594, Oct. 1995.

[8] Papandreou, N. & Antonakopoulos, T. (2005). A new computationally efficient discrete bit-loading algorithm for DMT applications. *IEEE Trans. Commun.*, vol.53, no.5, 785-789, May 2005.

[9] Chow, P. S., Cioffi, J. M. & Binham., J. A. C. (1995). A practical discrete multitone transceiver loading algorithm for data transmission over spectrally shaped channels. *IEEE Trans. Commun.*, vol. 43, 773-775, April 1995.

[10] Fischer, R. F. H. & Huber, J. B. (1996). A new loading algorithm for discrete multitone transmission. *Proc. of Conf. Rec. 1996 IEEE Global Telecommun. Conf. (GLOBECOM'96)*, London, 724-728, November 1996.

[11] Baccarelli, E., Fasano, A. & Biagi, M. (2002). Novel efficient Bit-loading algorithms for peak-energy-limited ADSL-type Multi-Carrier Systems. *IEEE Trans. on Signal Processing*, vol. 50, no.5, 1237-1247, May 2002.

[12] Medbo, P. Schramm. Channel Models for HIPERLAN/2. ETSI/BRAN document no. 3ERI085B, 1998.

In: Cognitive Radio
Editors: P.D. McGuire et al, pp. 139-202
ISBN: 978-1-60876-604-8

Chapter 6

COGNITIVE RADIO CONTROL: THE DISAPPEARING POLICY

Mikhail Smirnov
Fraunhofer FOKUS, Berlin

Abstract

The purpose of this chapter is to report very recent results on policy-based self-management towards self-adaptive and Cognitive Radio Systems (CRS). The focus will be not on self-* properties of cognitive radio systems but on self-learning policy and cognition; the chapter shows that self * properties appear as the result of cognition. The chapter will elaborate on radio domain-specific cognition model and on the role of context and policy in that model; specific emphasis will be given to policy multiplexing (this is where a policy disappears becoming embedded in to the fabric of other functionalities), process correctness (cross disciplinary approach based on information modelling and theory, control theory) as a part of assessment of CRS (like a "Turing test" for cognitive radio), and emerging design methodology of CRS.

Policy Multiplexing

Setting the Scene

Assuming that a reader is familiar with any reasonable definition of a policy, this section directly dives into the core problem of the entire chapter - policy multiplexing. For setting the scene one needs to compare the multiplexing phenomena in wired and wireless, in cognitive and non-cognitive. This will help to understand why the idea of cognitive radio is feasible, and why policies are so important for pragmatic use of cognitive algorithms in networking; it will also help to reveal the essential requirements the spectrum policies must satisfy in order to enable the learning, and by learning enabling better (cost-efficient, fair, robust, and versatile) communications. After that policy types and categories will be introduced in the next section on the background of evolution of policy concept in networking.

Multiplexing in Wired and Wireless

The vision of self-adaptive and cognitive network elements that promise to radically change our networking experience must be grounded in the reality and must take the path of a guided evolution; where the guidance is provided by thoughtful engineering with roots in fundamental research.

The need for cognition coupled with policy-based self-management in network elements for wireless systems was realised a decade ago [1, [2], [3]], however with the advent of autonomic computing and autonomic communications and with the on-going convergence of wired and wireless communications the cognitive networking is on the agenda of much broader as at the beginning community of designers. We use the term `advent' above in the non-theological meaning of `second coming' and demonstrate in the following sections that indeed self-, autonomic and cognitive properties of network elements were in the minds of early Internet designers. Do we have today sufficient guidance from theory, and from which one in order to address the current broad engineering scope?

The key difference between wired and wireless [packet switching] is in the nature of multiplexing. Wireless communications happen in essentially broadcast environment, which is largely being shaped by the communication itself; the multiplexing in wireless takes the form of positive and negative interference, and at certain level of communication intensity the cognition becomes a must due to the impossibility of even enumeration of all possible combinations [[4]].

Multiplexing in wired systems, even with multiple access (since it is a non-broadcast one) takes a form of a serialisation of service requests, which makes multiplexing controllable by means of e.g. capacity admission control, and even autonomically controllable by means of monitoring methods based on large deviation theory. However dynamic in-network service differentiation still remains a dream that can be achieved by means of self-adaptation and cognition. Cognition will enable autonomic multiplexing including much awaited service differentiation, or using the wording of Graca Cravalho [[5]] will enable "service creation per service element".

The Internet architecture evolved around the TCP/IP protocol stack that created the [wired] multiplexing environment, which in handling of all datagrams inherits the multiplexing properties of the stack. Let us concentrate on fairness property; fairness - conformity with rules or standards - is an important attribute of any service, therefore the fairness of multiplexing in relation to policy should always be evaluated. The TCP fairness is that of the FCFS processing discipline since all datagrams to be sent on the same network path will enter in sequence the same set of link buffers; the links constituting the path. When several TCP connections are sharing the same [sub-] path or even a single link, all of the connections experience similar datagram losses when the link (i.e. link buffer) is congested. The lost datagrams are obviously not acknowledged end-to-end by the receiving TCP instances as delivered, which triggers the sending TCP instances to reduce its current sending rate and subsequently to launch the slow-start algorithm: the slow start will try gradually to increase the reduced rate. It is important to mention that the rate reduction happens in a synchronised way between all affected TCP connections, which might lead to the under utilisation of the previously congested link. However, since the slow start of all affected connections is also synchronised the new congestion on the shared link might emerge[1] again, and so on. Figure 1.a shows that when the link is congested, say at time T_{cong} different connections sharing the link might detect the congestion, adjust their sending rate and launch the new slow start within certain interval, say $[T_{cong}, T_{adj}]$, which depends on the dynamics of the connections.

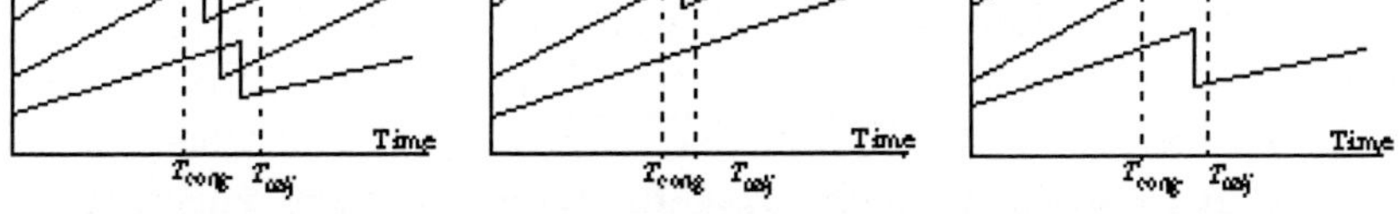

(a)Synchronised shift of TCP connections to slow start

(b)Shift of randomly selected TCP connection to slow start

(c) Shift of policy selected TCP connection to slow start

Figure 1. TCP behaviour under congestion: synchronous (a); controlled by RED (b); controlled by application policy (c)

[1] Under the reasonable assumptions that all previously affected connections will continue and that routing will not change during the described transition

These unfortunate synchronisation can be broken by the so called active buffer management schemes, with Random Early Detection (RED) being the most popular in-network solution. The RED [[6]] removes the synchronisation between slow downs of active connections by dropping random datagrams before the congestion occurs. For this RED monitors the statistical properties of a buffer. At the same time as [[7]]. states the simple "RED makes Quality of Service differentiation impossible", however, perhaps more fundamental issue of RED is that it breaks the fairness that is somehow preserved in the synchronous slow-downs. One might think that a simple enhancement can make RED if not fully QoS aware but at least sensitive to the traffic process as seen at the application level.

Suppose, every connection sharing the link is allowed to set a drop policy based on the criticality of the on-going application level activity; suppose also that RED is instructed to drop less critical rather than random datagrams, then one might expect better experience at the application level. Let us discuss this naive solution.

Typical outcome of the RED operation is shown on Figure 1.b; since the connection to be slowed down is selected by the algorithm randomly one might expect that the probability of datagram selection that belongs to the TCP connection with the highest sending rate among those that share the link is higher than for those connections that send at lower rates. Such an assumption is justified by the simple consideration that for a connection with the higher sending rate one might expect to find in the respective buffer larger amounts of datagrams belonging to this connection than to connections with lower sending rates. To reflect this dependency Figure 1.b shows in bold as slowed down the connection with the highest rate.

Figure 1.c shows what might happen when RED selection is guided by our naive policy: it is likely that the slowed down connection (shown in bold) will be that with the smaller sending rate. As a justification of this likelihood recall that low sending rate can be either due to the fact that this connection is fresh, or due to a recently experienced slow-down. The latter happens when our naive drop policy is not set for the connection, while the former happens when not enough time did elapse from the start of the connection, meaning that in general the connection is hardly in its critical[2] phase from the application viewpoint. The diagram also shows that selecting a low rate connection might result in longer adjustment times.

[2] Following the common utility logic: the longer lasts the connection the more work it brings to an end, hence value of the connection, and the criticality of its success are increasing

Based only on these pure speculative considerations one feels that something is wrong with our naive policy. Yet, further, any consistent attempt to engineer a system that is controlled by such policies will inevitably generate tough questions. The first question being: How to handle multiple and equi-valued drop policies? This might happen when all connections announce their datagrams as critical; in this case our imaginary policy-based RED will need to set priorities on equi-valued policies, which will require a lot of additional machinery to compute and to handle these priorities to preserve the four common service attributes - fairness, cost-efficiency, robustness and versatility [[8]]. As a result our naive and conceptually simple approach to drop policy requires not so simple secondary policies; because the above mentioned priorities of multiple policies can be seen as policies as well.

In the above it is assumed that our policy-based RED deploys at maximum one policy per connection; this policy is communicated to a RED by the sending application. However, even in the simplest case of a communication session there are two endpoints; critical states of application process can be experienced at receiving application as well. Consider, for example a financial transaction, in which the sending endpoint is that of a user, while the receiving is that of a financial server. The server can be entitled to set the receiving side policy, when the transaction enters a binding phase. In this case a RED instance can be faced with two policies for the same connection, and in general the possibility of conflicts between these two policies cannot be excluded. Obviously, yet another machinery needs to be added to detect and to resolve policy conflicts at run time.

Cases like the two above that complicate conceptually simple policy approach can be easily added when more features to be controlled by policies are considered. The TCP sending rate was the only feature under adaptation in our example, while practical networking demands to control adaptations that cater for mobility, security, dependability, quality of service (QoS) and the like. In wireless domain all these controlled adaptations must be preserved of course, however the radio itself demonstrates an explosion of possible adaptations that can be used by cognitive radio systems (CRS) built on flexible and programmable software defined radio (SDR) platforms. Following the example of a spectral awareness etiquette given in [[9]] Table 1 shows radio-specific adaptive features and required for this actions, mostly sensing.

Table 1. Spectral awareness etiquette features (after [[9]])

Adaptive feature	Performed CRS actions
Frequency	Using CSMA sensing CRS finds a frequency
TDMA	CRS finds an unused time slot in between a periodic user
Beam/Null steering	Transmitting CRS forms beam toward intended recipient, which forms Null toward interference sources
Bit loading	CRS Loads bits onto OFDM carriers based on SNR
Time-frequency holes	OFDM techniques where small spectral holes can be filled by one or a few carriers
Power control	Adaptive power control by e.g. shortest hop routing in ad hoc

Basically, policy-based control is a highly desired mechanism in networking; it controls feature adaptations and is attractive because policies are thought to be set at feature locations; this opens clear potential for radically distributed self-adaptations towards entire self-management of networks. However, as our naive policy example and its required complications demonstrate the associated mechanisms for handling of multiple heterogeneous and independently generated policies is very quickly becoming difficult to design and to maintain. Fundamentally, this difficulty is due to the need in the environment shared by all datagram flows to multiplex policies while preserving common service attributes at acceptable levels. Leaving rigorous definition of policy multiplexing to following sections let us consider now whether cognition, being implemented even by simplistic learning algorithms can help to keep the policy-based control simple yet powerful.

Towards Self-Learning Policy

Paul Baran is famous for his work on distributed systems; he was studying how the increase of the level of connectivity between autonomous network nodes helps to increase the entire survivability of packet switching network, i.e. to verify that such networks can survive significant topological changes. For this, he was in need of certain routing - fast and easy, which is since then known as "hot potato" routing. Interestingly, in this early design he did not separate routing from the rest of what we would now term networking middleware, and he termed “housekeeping”; he described a then unknown IP datagram like this - "Most of the message block would be reserved for whatever type data is to be transmitted, while the remainder would contain housekeeping information such as error detection and routing data" [[10]].

Though the centre of mass of Baran's work was on network structural properties he did find a very elegant co-optimization of routing and housekeeping, and it appears that he was the first to introduce in these settings not only policy but the cognition as well: "What is envisioned is a network of unmanned digital switches implementing a self-learning policy at each node so that overall traffic is effectively routed in a changing environment - without need for a central and possibly vulnerable control point... The network can be made rapidly responsive to the effects of destruction, repair, and transmission fades by a slight modification of the rules for computing the values on the handover number table[3]"[[11]].

Today, 45 years after that work we observe that policy is still in the focal point of our attention. Yet we are only demanding that "This evolution of policy, from 'just' a tool to tell if an action should be executed, to being intimately involved in how, where, when, and why configurations are changed and managed, is the next era of policy-based management" [[12]]. Obviously something went wrong with the Internet engineering; the process of organic growth made the Internet to become a patchwork of kingdoms [[13]] - policy interoperability and policy multiplexing are largely unsolved problems at the wholesale ISP interface i.e. between Autonomous Systems. As one of the pressing policy challenges of today in wireless the following did emerge (mainly within the cognitive networking domain, for example addressed in the E3 project[4]) - How to balance between autonomic decisions and centralised control by policies? Can we return to fundamentals and make networking right for a long-term evolution?

In the rest of this section let us attempt to design a policy mechanism for the above RED example, which mechanism shall exhibit nearly the same control behaviour as our naive policy but at the same time shall avoid all unwanted complications. This design example demonstrates that robust policy-based control is indeed possible, however policy as external to the controlled behaviour object disappears. Instead of purpose-made policy rule that traditionally is derived by refinement from a business goal and needs to be instantiated through network-specific and device-specific mappings, our policy is learned (inferred) by the controlled object from its operational situation.

Indeed, the inference needs extra resources - fortunately more modest and more generic than the above mentioned complications. As we know original RED monitors the statistical properties of a buffer; our aim would be to slightly modify this monitoring algorithm and couple it more tightly than in original design with RED decision (which datagram to drop) process.

---

[3] This is how Baran termed the routing table for the "hot potato" algorithm

[4] http://ict-e3.eu

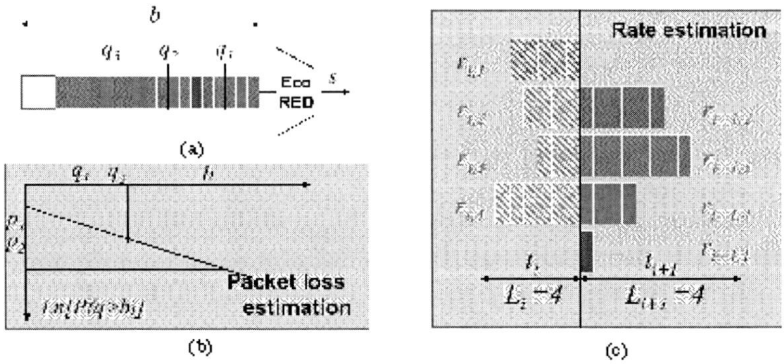

Figure 2. Fair (Eco-) RED: Self-generating and self-learning policy.

Conceptual model of our modification of RED design is that of an ecosystem of flows sharing the same buffer (link); each flow of datagrams is explicitly controlled end-to-end by the corresponding TCP congestion avoidance mechanism and implicitly - by the decision process of our EcoRED. Let us demand that drop decisions of EcoRED are fair.

We sketch here only the idea of the design and address the details in the section Learning Own Entropy. Single buffer is considered as logically being separated into a number of virtual buffers, one virtual buffer per active flow of datagrams Figure 2.a. We use fundamental result known from the probability (large deviations) theory: the large deviation rate-function (entropy) of bursty packet traffic can be estimated from measurements of traffic activity and it is linear (Figure 2.b). In other words, the entropy can be used to determine the bandwidth requirements of the traffic, which in turn can be used in fair decisions of a capacity admission control (CAC). However, typically RED is deployed in the backbone, where the CAC cannot be implemented. We use the following approach to bypass the issue: since RED will need eventually to drop a datagram, we can conceptually consider that would-be-dropped datagram does not pass the CAC. Basically, all datagrams within a buffer can be seen as not yet passed the CAC. Our EcoRED shall sample the buffer periodically (Figure 2.c), compute its own entropy, use the obtained information in making drop decisions - all this fully autonomously and without any visible policy though external observations of EcoRED behaviour should verify that it is in full conformance with the policy.

Policy Evolution

The section makes a very condensed overview of the evolution of policy concept; by explaining the roots and the progress of the policy concepts the section highlights important terms and properties. For example, when dealing with security policies the scalability property will be discussed; when dealing with configuration policies policy deployment modes will be discussed. Various flavours of policy taxonomy (ICL, NIST, DMTF, IETF, TMF, ACF, etc.) are compared with the attempt to semi-formally define necessary key requirements for policies that shall enable cognitive radio control in a truly autonomic - radically distributed and collaboration based - manner.

Introduction

As introduced in the previous section policy in networking, and more precisely in the Internet networking belongs to what Paul Baran has termed "network housekeeping". In a broad sense the housekeeping can be classified as in-network middleware. The term middleware in general information technology field refers to resources and services that are being so common in various applications that it appears cost-efficient to abstract these common things to a special layer that resides between the operating system and the applications.

In this respect it is interesting to recall a small debate that happened at WORDS workshop [[14]] in 2002. Observing that the middleware evolution gradually is abstracting more and more common services from applications, which services are also gradually but slowly becoming parts of operating systems, and that at the same time the evolution of applications that goes faster than that of OS creates yet another common services (read, a new middleware again between OS and applications), Prof. Michael Melliar-Smith of UCSB predicted "Middleware is going to be eaten by OS, OS is going to be eaten by HW...". In response Prof. Douglas Schmidt that time of DARPA arguably stated that "OS disappears, the only thing which will remain is middleware... the future of middleware is to become a part of an application; open source necessarily".

This debate did happen in the research community dealing with the application middleware; it's reasonable to ask ourselves How the network middleware is different? The emerging consensus in the networking community is that networks need a cross-layer network specific middleware. Cross-layer here means that entities residing at different levels of any protocol stack must be able to share information about values perceived by the respective entities at respective layers; network-specific needs more words to explain the difference. First, protocol entities - unlike various applications sharing the same OS and being facilitated by the same middleware - typically reside on heterogeneous operating platforms; it is widely recognised that heterogeneity of networking platforms will continue to grow. Second, the applications use middleware to re-use common services and resources while semantic interpretations of those remain application specific; contrary to applications the protocol entities need middleware to share common networking semantic. The former difference was captured long ago in IETF design and reads: "The world is a jungle in general, and the networking game contributes many animals" [[15]], while the latter difference tells us that these animals share the same set of values which are essentially conflicting. These shared values form the foundation of the network housekeeping platform, with policy mechanisms being the core components since policy is the embodiment of a value.

Bird's View on Network Housekeeping

As Figure 1 demonstrates the network housekeeping platform for a single autonomous system (AS) can be seen as being sandwiched between the two radically different types of interfaces - retail and wholesale. The retail interface faces the end-users of an AS, while the wholesale interface faces neighbour AS'es. A customer of an AS uses the retail interface by invoking a communication application, which in turn uses network interface (socket) to send and to receive payload data structured in respective protocol data units. Business relations between a customer and an ISP that owns the AS are regulated by an SLA, which typically for today's flat rate practice does not specify any service differentiation options; however with policy mechanisms deployed the much awaited service differentiation becomes possible. For this, an ISP must deploy a sort of OSS that will integrate the core components of the housekeeping platform, such as policy, AAA, directories, API's, resource management, discovery and retrieval tools, security and QoS mechanisms. It should be noted that for clear reasons ISP's build their networks from equipment originated from different manufacturers, thus vendor-specific support for OSS functions is typically limited.

The fundamental difference between the retail and wholesale interfaces - and as a result between the two types of markets - is their radically different ability to support by in-network features handling of payload data at all layers of the hierarchy "application - network - mechanism". An ISP has full control over its retail market but only within the connectivity range that does not cross the AS boundary. These in-network features do not inter-work across the AS:AS boundary, though edge routers even coming from different vendors are able to support such inter-working if conforming to IETF standards.

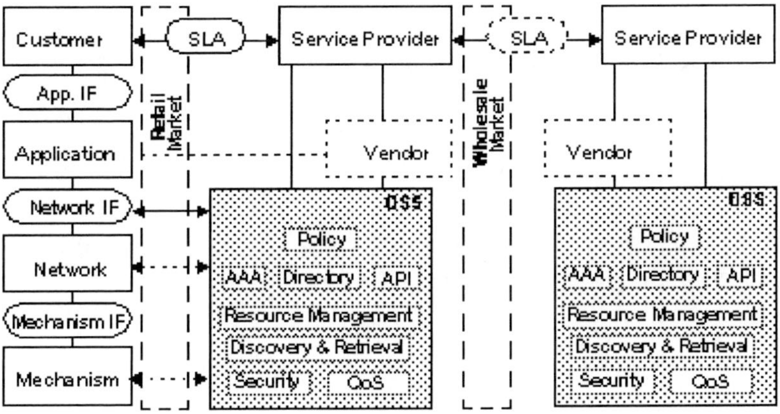

Figure 3. Network Hose keeping reference model (IF - interface; SLA - service level agreement; OSS - operation support system; AAA - authentication, authorisation and accounting API - application programming interface; QoS - quality of service)

The issue is well known and is addressed by the engineering community, especially by the IETF[5] and by the TMF[6]. The IETF is historically concerned with the end-to-end communication and its properties, while the TMF's goal is to help network operators to do their business. These different viewpoints are clearly reflected in the respective approaches to solve the issue of the wholesale market.

The TMF approach is known as next generation OSS (NGOSS) [[16]], which is shown in Figure 4. as a replacement of OSS, but not only. The NGOSS attempts to facilitate feature inter-working across the domain boundaries yet keeping the domains loosely coupled. The main facilitator of the inter-operation of OSS'es is known as Shared information/Data Model (SID) that is commonly used by the inter-working systems to orchestrate control over a connection spanning both systems. Each system runs its own workflow engine that serves incoming application requests (the bold line connecting application interface and the workflow engine on Figure 4); domains are assumed to collaborate using shared SID semantic. The SID is also used to configure features themselves, in

[5] Internet Engineering Task Force, http://www.ietf.org
[6] Telemanagement Forum, http://tmforum.org

particular those management components that are responsible for service Fulfilment, Assurance, and Billing (FAB). Notably, the legacy FAB components can be re-used if wrapped appropriately to conform to the NGOSS's Common Communication Infrastructure - sort of information bus interfacing wrapped FAB components and the rest of NGOSS.

The IETF documented its position on Internet housekeeping in [[17]], which basically states that the required housekeeping has no particular structure or hierarchy; in a truly Internet spirit the housekeeping is described as a set of components - resources and services - that can be used by any network entities. Perhaps, housekeeping as such is not the primary concern of the IETF however end-to-end service control especially in need of service differentiation are in the focus of the IETF work for a couple of decades. That is why, the key component of the housekeeping – the policy has received significant attention.

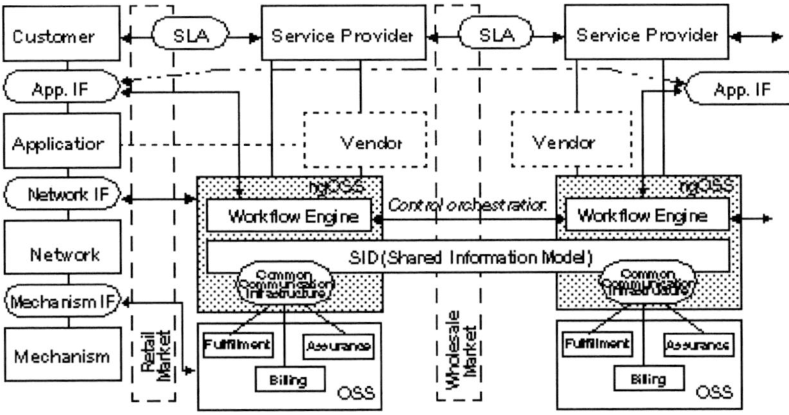

Figure 4. NGOSS within the Housekeeping Reference Model.

The IETF Policy Framework working group defined policy framework definition language, policy architectural model, associated terminology, and policy models for both signalled QoS and provisioned QoS. The main principle is that of policy refinement - step by step detalisation of a business goal reflected in a SLA to a low-level device implementation mechanisms with the use of policy model. The following four policy mechanisms were defined:

- access control objects
- service management objects
- resources
- configuration of mechanisms.

If TMF approach is to enable collaboration of loosely coupled NGOSS'es by leaving somehow end-to-end control out of primary attention, the IETF approach pays little attention to the inter-domain collaboration assuming that service differentiation mechanisms deployed end-to-end in all network elements that handle the service traffic along the path will automatically make domains to inter-work. That is why the main design framework for policy engineering within the IETF was so called policy domain presented on Figure 5. The two main entities are assumed in this policy model - policy decision point (PDP) and policy enforcement point (PEP). The PDP is the entity of a policy service directly facing the network; the policy service includes all needed machinery to create and manage abstract and refined policies. The PEP is co-located with every network element that needs to be controlled by policies.

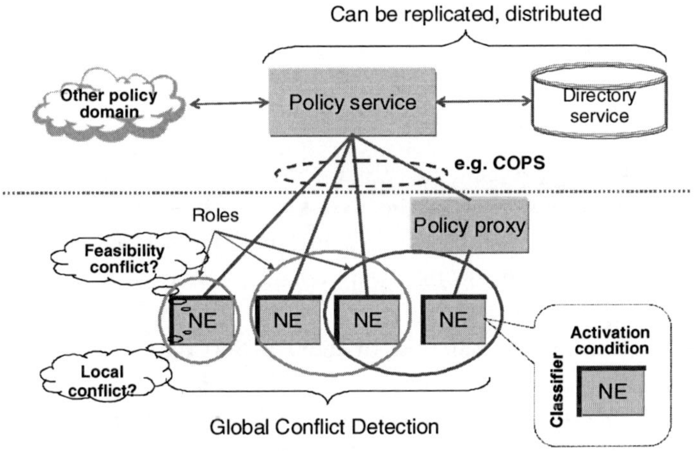

After QoS Policy Framework Architecture, draft-sgai-policy-framework-00.txt
NE –Network Element; COPS – Common Open Policy Service

Figure 5. IETF Policy Domain (NE - Network Element; COPS - Common Open Policy Service [RFC2748]).

The framework does not assume that all enforced policies are conflict-free; it recognises that certain types of conflicts can be detected only at PEP and only after the enforcement of policies, for example so called feasibility conflicts, in which case the PEP is erroneously associated with a network element that does not have a functionality required by the enforced policy. The framework allows to group policies based on common activation conditions; in this case network elements enforcing group policies are said to be assigned certain roles.

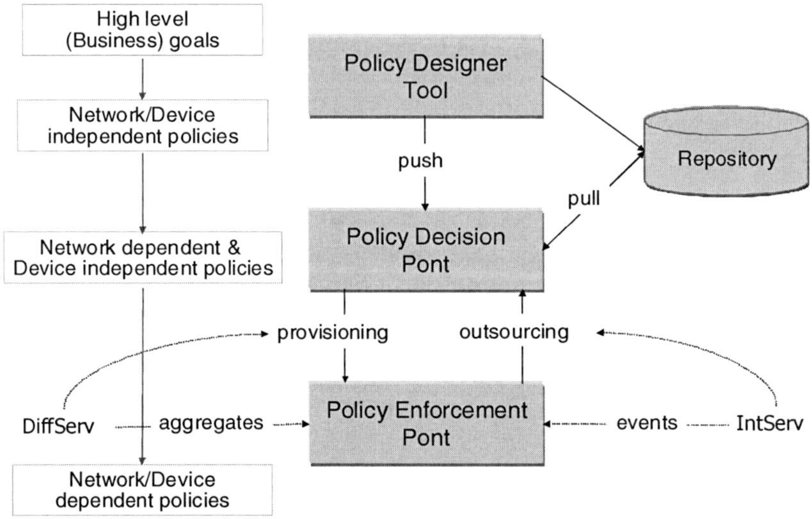

Figure 6. IETF policy provisioning vs policy outsourcing.

The greatest achievement of IETF policy work is the instantiation of developed policy framework for the two QoS architectures engineered by the IETF - integrated services (IntServ) and differentiated services (DiffServ). In the case of IntServ the individual QoS is signalled; the network element receiving a QoS request (event on Figure 6) outsources the decision to PDP, which returns properly refined policy to that NE. In the DiffServ case the QoS requests are aggregated which makes feasible to pre-install (provisioning on Figure 6*)* properly refined policies on all network elements of a DiffServ domain.

In these two examples - TMF and IETF policy framework - we observe that policy mechanisms are strongly depend on information modelling. An information model (IM) in software engineering is defined in [[18]] as representation of concepts, relationships, constraints, rules, and operations to specify data semantics for a chosen domain of discourse. It can provide shared, stable, and organized structure of information requirements for the domain context. The IETF adopted the Common Information Model (CIM) developed by the DMTF[7], while TMF has developed its own SID model. The reliance on an IM is fundamentally important for any policy-based system since it allows network elements to exchange their values; this is possible because the values are based on common semantic captured in an IM.

---

[7] Distributed Management Task Force, Inc., http://www.dmtf.org/home

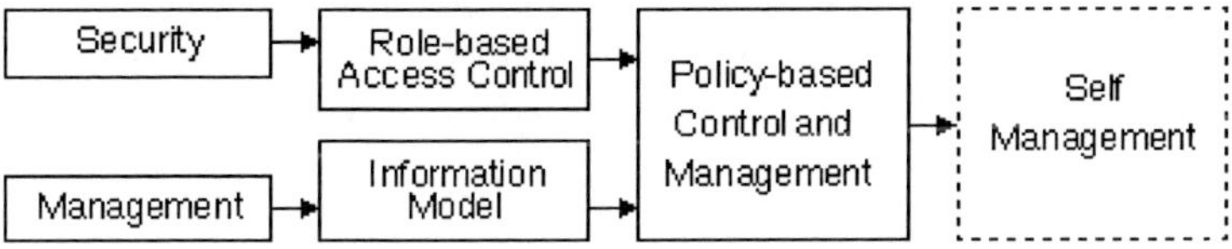

Figure 7. Policy Evolution.

The transition from management stovepiping to IM based systems with shared semantic is shown at the bottom of Figure 7 as one - management (mainly device configuration) - part of policy evolution, while the other part is the of evolution of security policy. If management policy evolves due to the need to increase the scope of shared semantic, the security policy (mainly access control (AC) to resources, devices, functions) evolves due to the need to increase the scalability, while the semantic of security policy cannot be shared since it must be strictly individual decision about the access.

The access control framework distinguishes between subjects and objects; subjects request invocation of actions[8] on objects. The AC is used in many areas, such as networked file systems or machines, firewall configuration, organisations (banks, financial brokers, health care institutions, etc.), B2B and e-commerce marketplaces, but also at a very low level, or even in hardware in datagram processing workflows in hosts and routers. There are many forms of AC, for example one distinguishes between implicit and explicit access control decisions, however since scalability is the major reason of AC to evolve it makes sense to trace this evolution.

The simplest possible access control is implemented as an access matrix having, say subjects as rows and objects as columns and permissions in the respective cells of the matrix. Permissions for subjects to request actions on objects are set as direct relation between them as shown on Figure 8.a. Obviously, this type of AC does not scale well since the size of the matrix grows as the product of the amounts of subjects and objects. Though the access matrix is conflict free it does not scale and is static. Observing that both sets - subjects and objects - are considered as unstructured ones, it is possible to improve the access matrix by taking the structure of both sets into consideration. The widely known extension of access matrix that takes into account the structure of subjects is Role-Based Access Control (RBAC).

[8] Obviously, a resource usage can be also termed an action

The simple RBAC uses access matrix to relate by permissions subject roles to objects, additional matrix is used to relate subjects to their roles - this is where the structure of the set of subjects is taken into account. The RBAC scheme is more flexible since the assignment of subjects to roles is independent from the assignment of role's permissions to objects as shown on Figure 8.b; it has also better scalability since the complexity grows as the sum of two products - the product of the amount of roles to the amount of subjects and to the amount of objects. Even the simplest RBAC has two additional nice properties: it allows dynamic assignments of roles to subjects and some sort of self-assignments of permissions. The granularity of the former is theoretically per session, while the latter is achieved by the permission propagation rules on the graph representing the role structure.

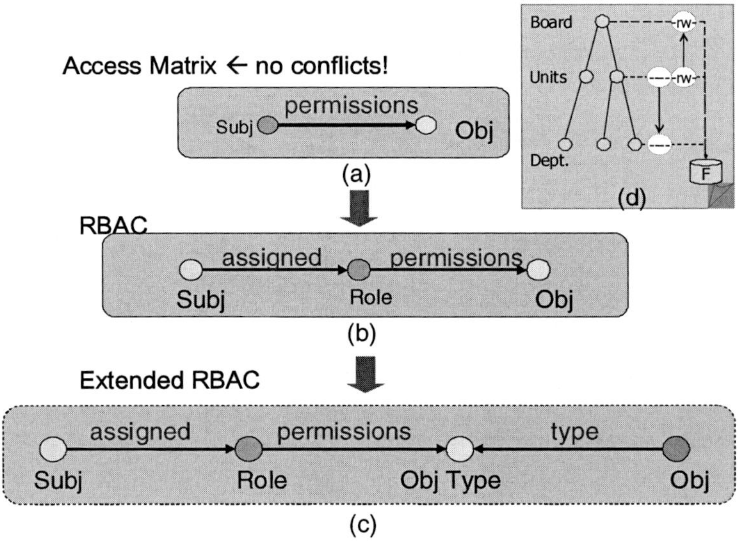

Figure 8. Evolution of Access Control: (a) - Access Matrix; (b) - Role-based access control; (c) - extended RBAC: (d) - propagation of permissions in RBAC.

Figure 8.d shows that if the structure of the set of subjects is a tree, in which the highest level role is that of a Board member, the next level is for the Unit head role, and the bottom is for the Department head role, then positive permissions (for example read-write permission on certain files) will propagate upstream the hierarchy, while negative permissions will propagate downstream. However, the RBAC is not conflict free: multiple dynamic roles assignments are possible, as well permissions may have multiple inheritances.

Obviously, simple RBAC can be extended based on the structure of the set of objects. As Figure 8.c shows one possible way to extend RBAC is to define permissions between subject roles and object types, this will have two additional matrices in the background - for mapping of subjects to roles and for mapping of objects to their types. For completeness, the below summarises some other possible extensions of RBAC:

- Temporal RBAC will use time-dependent permissions;
- Team-based RBAC will group subjects in Unix-like groups, which is useful for collaborative environments;
- Domain-based RBAC will group objects into domains with similar permissions as implemented in Ponder language;
- Context-based RBAC will facilitate context-aware decisions, where context can be time, location, environments, etc.

As some intermediate conclusion from this quick survey of policy evolution shown as the left part of Figure 7. let us repeat major drivers of this evolution. Management policies evolve by means of their reliance on common information model, which assures common semantic shared by a larger scope of network elements; security policies evolve by means of various mappings exploiting the structure of subjects and objects. It is the right time to ask ourselves the question: Aren't those two means actually the two sides of the same coin? Indeed they are. A good information model shall reflect subject roles, object types, structures of the both sets, including relevant grouping (domains) as well as relevant contexts, as well as their relations. Because of this commonality between the policy evolution drivers the right part of Figure 7 shows both evolutionary paths converging to policy-based control and management.

Indeed the ISO standard captures the telecommunication network management tasks as FCAPS, which already include security as a part of management. However, FCAPS is a model, while sadly in practice the three things that are being overlooked most often during the functional design are: security, manageability and testing. Virgil Gligor puts it as follows: "In the history of computing there has often been a 10 or more year gap between the use of technology and the addressing of security issues that arise from it" [[19]]. Already in 2002 John Strassner stressed in [[20]] the need for "... evolution of policy, from 'just' a tool to tell if an action should be executed, to being intimately involved in how, where, when, and why configurations are changed and managed, is the next era of policy-based management." With this in mind let us observe what are the essential features of policy system that are required to facilitate not only converged control and management empowered by a shared information model, but also shall facilitate further evolution of that system towards self-management of network infrastructures.

The first required feature is that of computational completeness of a policy system, and the second one is the inclusion of logic. The computational completeness is required from a policy system itself, while the logic can be embedded in the foundation of a policy system - in the underlying information model. The rest of this section covers policy approaches that have these desired features - policy work (implemented in Ponder language) by Prof. Moris Sloman[9] and the IM (also known as DENng, which stands for the next generation of Directory Enabled Networking) work by Prof. John Strassner[10].

According to M. Sloman the following two types of policies must be present in a policy system: authorisation policy (AP) and obligation policy (OP). The AP specifies what actions a subject is permitted or forbidden to perform on a set of target objects, while the OP specifies what actions must be performed by a subject on a target objects. Note, that obligation policy is often thought of as redundant, and negative obligation is often regarded as hard to understand. However, from the computational completeness viewpoint negative obligation plays the same role as inhibitor arcs in Petri net models, which makes the language of Petri nets extended with inhibitor arcs computationally complete in the Turing sense.

Syntactically, the OP and AP are indistinguishable: Figure 9 shows that both types of policies are specified with the same set of attributes, policy type being one of those. From the syntax of thus specified policy it is immediately clear that it is very easy to make mistakes in policy specification that will lead to various types of conflicts (feasibility, local/global; self/non-self; modality/semantic; etc.). Though policy languages[11] such as Ponder2 and its higher level configurer PonderTalk make feasible automated creation of policies, still conflicts are possible. M. Sloman provides the following generic recommendations for resolving policy conflicts:

- Sub-domain policy overrides domain policy;
- Negative modality overrides positive modality;
- Recent policy overrides older one;
- Short term policy overrides long term one.

[9] Professor at Imperial College London, UK, http://www-dse.doc.ic.ac.uk/cgi-bin/moin.cgi/mss
[10] Professor at Wateford Institute of Technology, Ireland, http://www.tssg.org/
[11] Available at http://www.ponder2.net/

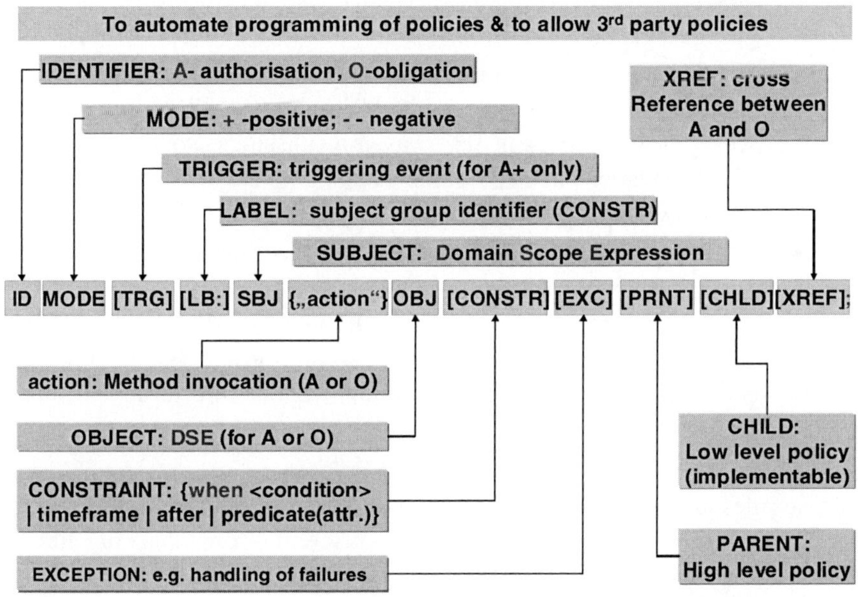

Figure 9. Policy specification in Ponder.

These recommendations are in essence rules that when applied to conflicting policies decide in favour of more reasonable one, based in this case on common (human) logic; these rules are meta-policies, i.e. policies applied to policies. This way, already Sloman's policy framework has some logic though this logic has a meta-nature, has implicit flavour of recommendations. Is it possible, as postulated above to have all needed logic as part of a policy system?

But, before we should ask ourselves - Do we know all needed logic? As shown in Table 1, the radio link layer explodes the amount of possible adaptations, yet it expands the amount of possible conflicts of various types. It seems that the earliest attempt to describe these meta-rules for radio ecosystem are the "kindergarten rules" by Paul Baran; following the [21] they are quoted below.

- Keep away from the big bullies in the playground. (Avoid the strongest signals.)
- Share your toys. (Minimize your transmitted power. Use the shortest hop distances feasible. Minimize average power density per Hertz.)
- If you have nothing to say, keep quiet.

- Don't pick on the big kids. (Don't step on strong signals. You're going to get clobbered.)
- If you feel you absolutely must beat up somebody, be sure to pick someone smaller than yourself. (Now this is a less obvious one, as weak signals represent far away transmissions; so your signals will likely be attenuated the same amount in the reverse direction and probably not cause significant interference.)
- Don't get too close to your neighbour. Even the weakest signals are very strong when they are shouted in your ear.
- Lastly, don't be a cry baby. (If you insist on using obsolete technology that is highly sensitive to interfering signals, don't expect much sympathy when you complain about interfering signals in a shared band.)

These rules are very properly termed since learning abilities of children in the kindergarten age are known to be very different from that of adults. A psychologist or ethologist would describe these abilities as imprinting, i.e. "phase sensitive learning (learning occurring at a particular age or a particular life stage) that is rapid and apparently independent of the consequences of behaviour" (quoted from [[22]]). Cognitive radio systems are expected to be aware of the consequences of their behaviour; therefore the imprinting - acceptable at the training phase - should be substituted by very dependable learning at the real world operation. That is why the logic embedded into a policy system of cognitive radio, or in its underlying information model must facilitate transformation of predicates according to the shared semantic of their IM. DENng appears to be the most advanced IM to the moment of this writing; the rest of this overview section shall summarise its useful features.

DENng is the information model enabling the design and the desired operation of an autonomic network and/or autonomic network elements. The model is positioned by the ACF[12] and by the model's chief architect, the ACF chair Prof. John Strassner as the major standardisation target of the ACF. The DENng model is evolving over time, having its roots in DMTF's CIM and in TMF's SID though it is radically departing from both CIM and SID. Using J. Strassner's wording the DENng helps to overcome the UML limitations, in particular the fact that "MOF is not formal" [[23]].

---

[12] Autonomic Communications Forum, http://www.autonomic-communications-forum.org

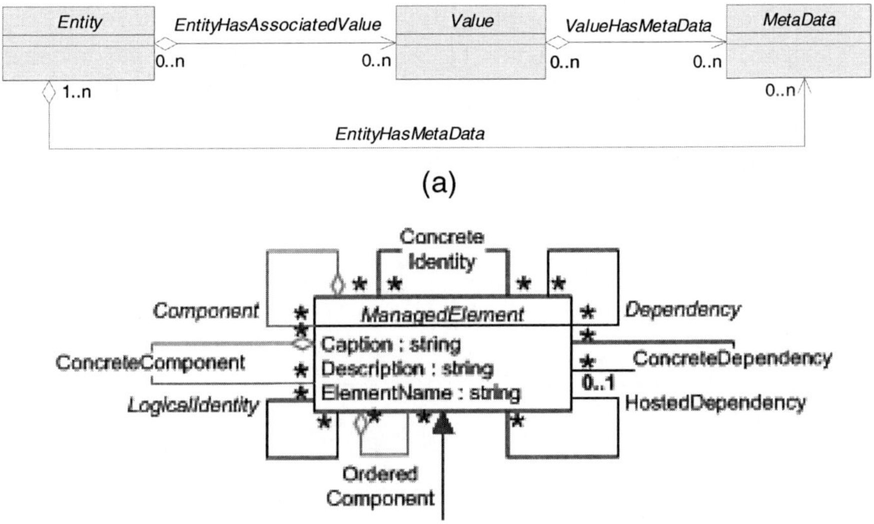

(b)

Figure 10. Comparison of top levels of DENng (a) and CIM (b) (after [[23]]).

Abstractions are of the key importance to the field of information modelling. These abstractions in DENng take the form of aspects; the aspects can be logical resources, contexts and roles. DENng consistently considers as aspects the answers to the questions Who (a role to configure); How (which method), Why (under which reason), What (which "thing"), When and Where.

Besides the fictional Root node the top level of the DENng model is comprised of the three classes - Entity, Value, Meta-data (see Figure 10.a), which is in radical difference to MOF, which would have at this level Attribute, Class, and Instance, and even more different from the CIM, which top level is comprised of a ManagedEntity (see Figure 10.b), which has 8 recursive relations that are all inherited by all other elements

Patterns are being heavily used already in SID, however this usage appears to be more consistent in DENng, which distinguishes between the following patterns:

Patterns = {behavioural, structural, creational, architectural}.

Consequently the modelling proceeds is using the patterns like "Role-Object", "Composition", "Policy", "Management" and the like; notably the Policy pattern allows to engineer policy-based managed system (PBMS) during the design phase. The interoperability (~compliance) is achieved via the standardised set of core elements. As shown in Figure 11 the policy pattern is so generic that is applicable for various uses. Notably, the "XYDetails" component that is being responsible for required mappings can be in principle a placeholder for the transformation of predicates mentioned above based on learning capabilities of a CRS.

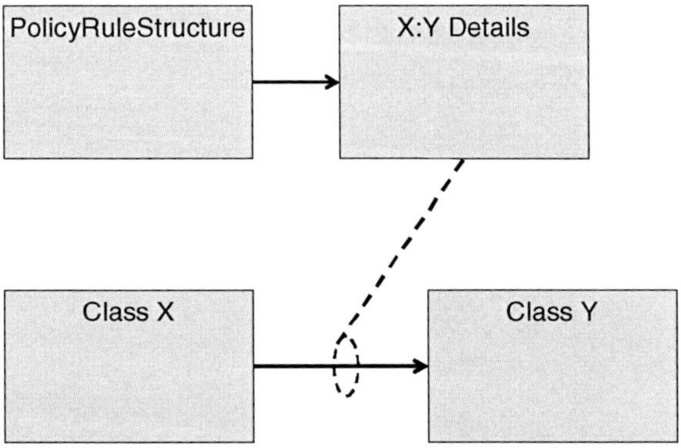

Figure 11. DENng Policy Pattern (after [[23]]).

Additionally to the conventional modelling of a policy rule in the Event-Condition-Action (ECA) form, the DENng considers also Goal policies and Utility policies; consequently the very high level model of a policy becomes:

Policy = < ECA, Subject, Object, Category>,

where ECA is a conventional rule form; Subject and Object are correspondingly policy subject and policy target object, and policy category defines the level of the so called policy continuum. The subject and object are conceptually similar to those of the M. Sloman policy model (considering both positive and negative modalities for obligation[13] policies that are set on subject, and both positive and negative modalities for authorisation policies that are set on the target object), while policy category is a novel concept. In DENng policy continuum is the complete refinement chain of business goal into [a set of] fully qualified technical policies, thus the Category defines the level of this refinement. This representation of policy appears to be very useful in the above mentioned PBMS engineering.

[13] In DENng obligation policies are termed deontic policies

The DENng features strong dependencies on the classification theory, thus the modelling logic is the one of building hierarchies. Note, however that the resulting model of a communication system does not have conventional layers (as e.g. in ISO OSI RM), the modelling hierarchies are somewhat orthogonal to the conventional layers of a protocol stacks.

The top level Entity in DENng appears to be decomposed into

{Event, UnmanagedObject, ManagedObject},

where ManagedObject is always associated with the so called Behavioural Aspect of DENng model; a location can be an example of an UnmanagedObject.

Behavioural Aspect of a model can be further decomposed as

{Policy, Context, Interaction},

which is another remarkable feature of the model. since Interaction serves the purpose of modelling of communication in general; for example its instance of Business Interaction will have then an SLA as a model.

The IETF policy model assumes that each policy domain has a single (or also proxy) PDP and a number of PEPs. Contrary to that the DENng allows multiple PDPs (which in general much closer corresponds to the reality), and considers PEP as an abstraction, which is further decomposed into PolicyExecution and PolicyVerification. Following the policy continuum concept the goal policy can be represented at different levels; the vertical interaction between the four points above can proceed as a feedback or as a feed forward thus catering for control loops.

DENng strongly supports various types of composition, which is critically important for self-managed networks. A composition is being defined by at least one physical resource and at least one logical resource (see an example on Figure 12). This way, only logical and only physical resources do not form a composition in the DENng terms. The hierarchy of logical resources, which are also managed transmission entities is given by

Logical *Resource = {Logical Device* | *Network Address* | *Device* Interface}

Further Device Interface is modelled as media independent and media specific. It was also stated that "QoS are services", and more precisely Customer Facing Services.

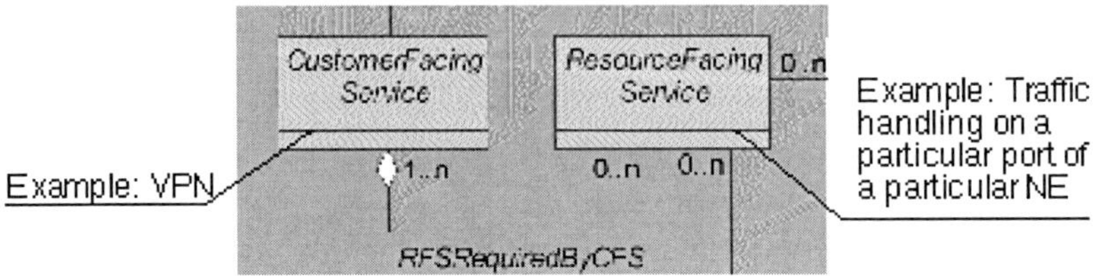

Figure 12. Example of a composition (after See John Strassner, "Introduction to DENng for PanLab II", ACF, 2008-2009, tutorial given 21.01.2009 in Fraunhofer FOKUS.).

Main Findings

Concluding this overview section let us summarise the main findings from the evolution of policy and its enablers that are important for the rest of the chapter.

Policy belongs to network housekeeping. Contrary to application level middleware that facilitates the re-use of services and resources and leaves semantic interpretations of these uses to applications, the network housekeeping platform facilitates the re-use of semantic that is common to all network elements in the following two senses. First, because this is the semantic of datagram handling and it is being created by network elements themselves, and second, because this is the semantic of underlying information model, to which all network elements do conform. If the former cause is true in today's networks, especially for packet radio networks, the latter cause is a fortunate scenario for a near-term future, when all software code in the SDR enabling CRS will be generated based on the same IM. The policy within the housekeeping platform is the embodiment of value; the values being different for different stake holders.

The two very distinct values within the interconnected TCP/IP networks are those known as end-to-end (end-user value) and the domain (network operator value) ones. Housekeeping platforms designed to protect the two types of values are not really integrated with each other. This results in known issues with network dependability for business use, mainly due to the lack of inter-domain solutions for network elements configuration management and security management, and of course due to the lack of inter-domain policy interoperability.

Over time both configuration and security policy evolve, however differently.

Finally, the computational completeness and logic are important ingridients of a policy system.

Notation for Self-Learning Policy

The notation that should be used throughout the chapter is novel and is specifically invented to meet the requirements (cognition, policy multiplexing) of the study reported here.

Using the above notation the section builds a model of self-learning policy - the key to cognition algorithms and to the robust control by policies. The model and the notation capabilities are verified by their application to classical problems, such as remote method invocation dilemma, control robustness by policies, optimization of heterogeneous wireless networks.

Why Notation?

The previous section advocates IM techniques as the underlying mechanism for designing policy systems and network elements that can be controlled by policies. Why yet another formalism is needed? The short answer is: policy notation must enable a designer with the tools for reasoning about policies, especially about policy multiplexing.

Policy becomes the first class citizen in the IT world; as an object in this world a policy can be attributed by all relevant aspects and very quickly becomes a very complex object! This is true for the policy specification in Ponder (see Figure 9) as well as for any other consistent attempt, like presented on Figure 13.a

Similar to Petri Nets that are excellent for reasoning about hard problems of concurrence but are rarely used for the design of concurrent systems, the notation proposed here must empower a designer with the ability to analyse hard problems in policy, and after the right solution is obtained the design can proceed to the specification phase, in which IM tools are more useful.

What are these hard problems? In the area of CRS the focus should be neither on learning nor on adaptation per se, but on domain-specific problem solving. Both learning and adaptation are internal to a CRS - we can't observe them, hence we can't evaluate (assess) them. While externally defined, observable (i.e. objective) situations might form a problem for a CRS, which problem, when solved (or not solved) can be observed, and hence the CRS's ability to solve problems can be assessed.

To give an example of invisible learning is easy - just think of a system that accumulates knowledge but never applies it to problem solving - but here is an example (see Figure 13.b) of an invisible adaptation. Imagine a CRS that monitors relayed traffic, when the traffic rate is less than the threshold R_l the CRS does nothing, when the rate is more than R_1 the CRS launches internal adaptation process that is preparing to handle future traffic if its rate would become more than R_2. The lower part of the Figure 13.b shows that if the rate never exceeds the R_2 we would never observe new behaviour but the adaptation and learning were done.

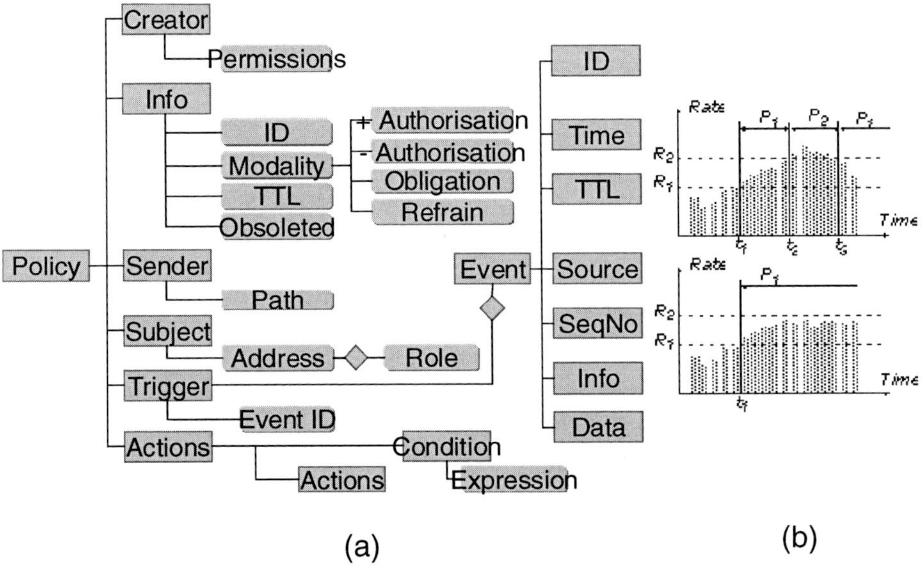

Figure 13. (a): Policy Specification Structure (after [[24]]) (b): invisible policy adaptation.

This simple example takes into account system's internal resources that are required to handle different situations by different policy rule sets - P_1 and P_2 respectively. In the existing body of policy work these resources are rarely taken into account, perhaps because those are regarded implementation specific. However, when applying policy in the area of CRS one hardly can neglect these internal resources, or, in other words the overhead induced by policy-based control[14].

The main reason is that CRS must deploy adaptive radio frequency sensing, and that such sensing is multifaceted (see Table 1). It will be inefficient in every aspect for a CRS to try to deploy concurrently all possible types of sensing.

[14] In conformance with the policy evolution roadmap outlined in the section on policy evolution this section assumes that in self-managed networks control and management are indistinguishable

Contrary to the above the policy-based reconfiguration of sensing appears very attractive. Not only in wireless - any networking that requires self-adaptation will need, as Figure 13.b demonstrates several sets[15] of policies to be enforced when the respective situation is sensed/detected.

The need to account for internal resources used for policy reconfiguration is an important requirement for the intended policy notation. This translates into the need to extend the notion of conventional policy domain. Without loss of generality we assume that each radio access technology (RAT) is a policy domain, meaning that all [behaviours of all] entities using particular RAT are configured by the same set of policies.

Since so far mentioned entities are either subjects or objects we need to see whether other types of entities are redundant for the notation. Conventionally in existing policy-based systems the policy middleware (sometimes termed policy agent[16]) is the entity that by assumption knows the policies of all subjects and the actions available on all objects; the policy agent in such designs serves as a communicator and simultaneously as the policy admission controller between subjects and objects. In the attempt to eliminate the policy agent and to cater for the design of systems that are natively controlled by policies we advocate that the composition principle is useful. Firstly, we concentrate on policy semantics as seen by a policy agent (a middleware viewpoint), and secondly we interpret the same policies from the viewpoint of a system, in which subjects are composed with objects.

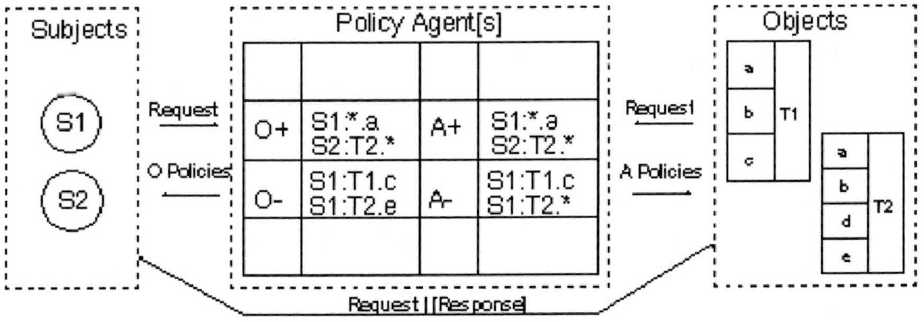

Figure 14. Middleware semantics of policies.

[15] It is not required that these sets are non-overlapping; moreover it is not required that these sets do exist permanently - cognitive system is able to build them dynamically when needed.

[16] PDP in the IETF policy framework is a good example of policy agent

Consider an abstract example of two subjects (S) and two target objects (T), and a policy agent that holds both types of policy as shown on Figure 14. Subjects are represented by circles; target objects are represented as boxes assumed to host a number of methods - would be invoked actions when requested by subjects - methods/actions are denoted by small characters inside smaller boxes within targets. The policy agent groups obligation and authorisation policies per modality, denoted as `+' or `-'. A subject either requests policy agent and gets in return (outsourcing) or gets via push (provisioning) its obligation policies; similar exchange happens between objects and policy agent. After that configuration phase is done subjects are able to request actions from the target agents.

Even this simple example demonstrates that the configuration of policy agent is a complex task. Figure 14 uses the following syntax:

<PolicyType>{+ | -}<Subject>:<TargetObject>.<Action>,

assuming common use of wildcards. Like in real life the set of policies in our example is not complete in the two senses. First, not all relations between the two sets are defined, e.g. no obligation is specified for S2 with regard to actions of T1. Second, among defined relations not all match in pairs, e.g. A-S1.T2.* is only partially matched by O-S1.T2.e. This lack of completeness is perhaps not that dangerous in systems with static relations between the two sets, however CRS are expected to be self-managed (at least partially), which requires dynamic assignment of Subject and Object roles. In these dynamic settings the incomplete definitions of policies or lack of meta-rules (like permission propagation) make middleware based approach not feasible.

From the theoretical computer science viewpoint the semantic presented in Figure 14 is that of a glue code based on abstract data types (ADT) and is usually termed rendezvous semantic, which can be explained with the following quote from [[25]]: "The method invocation semantics of object oriented message passing implies a rather tight semantic coupling between the caller and callee pairs of objects. By this semantics, if an object *c* sends a message *m(p)* to another object *e*, then *c* is invoking the method *m* of *e* with the actual parameters *p*. For this to happen: i. *c* must know (how to find) *e*; ii. *c* must know the syntax and the semantics of the method *m* of *e*; iii. *e* must (pretend to) perform the activated method *m* on parameters *p*, and return its result to *c* upon its completion (the pretense refers to when *e* delegates the actual execution of *m* to a third object); and *c* typically suspends between its sending of *m* and the receiving of its (perhaps null) result."

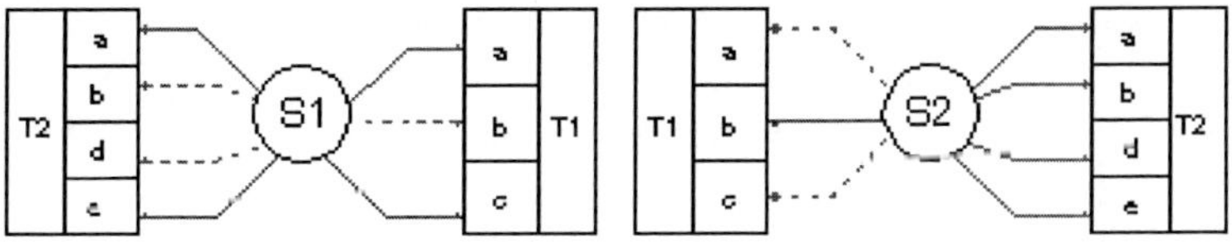

Figure 15. Policy semantic based on composition.

To explain the policy semantics from the composition viewpoint we apply the zero knowledge reasoning to both types of entities that appear in a composition - to subjects and to objects. The zero knowledge breaks the rendezvous semantic; in particular, all but the first must requirements from the above are relaxed. We substitute the first must know (how to find) by an assumption that all entities in a composed system are fully connected (or, can discover such a connectivity) and can pass arbitrary messages to each other. These arbitrary messages will be those that each entity can form and send; without loss of generality we can assume that these messages are those that subject and objects are configured to form and to send under certain conditions. Of course, an action of passing a particular message triggered by some conditions can be seen as a policy; however for the moment we intentionally forget about this. Instead, we imagine that entities pass messages (those that they can pass) unconditionally and for a reasonably long period of time; we also assume that all sent and all received messages are timestamped and stored in respective received and sent buffers. Eventually each entity will pass all messages it can send and will receive all possible sent to it messages in the composed system.

When this happens, we require each entity to start relating (matching) all sent and received messages to the entity's configuration conditions. During this hypothetical matching process the subjects of the composed system will observe that all sent messages match the configuration and some of the received messages also match the configuration; the objects will observe that some of the sent and some of the received match the configuration. This reflects the fundamental difference between the two roles - subjects and objects; in our mental exercise the subjects discover the obligation policies (all sent messages are configured as must request) and the objects discover the authorisation policies (some sent messages are matching some incoming requests that may be answered), by which they were configured. Further analysis of received and sent buffers will allow subjects to infer per obligation policy the sets of objects that have positive and negative authorisation for the respective requests; by the similar analysis objects may infer per authorisation policy the sets of subjects that may and may not request respective actions.

We argue that in a correctly composed system the subjects act as clients, while the objects act as servers; the objects' configuration serves the purpose of a safeguard, while subjects' configurations are behaviour definitions. This argument facilitates the design of systems, in which subjects and objects can be in principle dynamically composed. Note also that composition shown on Figure 15 exhibits one additional nice property, namely system's self-configuration ability, which is somehow similar to permission propagation schemes in RBAC. Since policy agent configuration was not complete, we assumed the same incomplete configuration to be implemented on subjects and objects; the bold lines on Figure 15. show that composition relations (dark green as positive and red as negative) that were discovered when the entities followed the configured policies, those shown in dashed lines could be eventually discovered without particular configuration.

Thus, reasoning about policies in terms of compositions of subject and objects is useful; it is also very generic - policies are thus somehow similar to chemical bonds, and if one continues this analogy then one comes logically to the ideas of policy strength (how tightly it couples the two entities), and policy energy, which we shall explore later by showing the relation between the amount of system's internal resources needed to learn certain policy and the entropy of that system and its environment. However, before that, we need to describe means that are better than the above zero knowledge discovery mechanism, with which we can design systems that deliver controllable and predictable behaviour. For these means we need certain facilitation from the control theory.

Recently published formulation of an unsolved control theoretical problem - decentralised control with communication between controllers [[26]]- clearly states that "control of traffic on the Internet is a concrete example" of that unsolved problem. The problem reads: to synthesize *R* controllers and a communication protocol for each directed tuple of controllers, such that when the controllers all use their received communications the control objective is met as well as possible. Notably, a communication protocol is called a communication constraint in [26]; it concludes that the basic underlying problem is "What information of a controller is so essential in regard to a control purpose that it has to be communicated to other controllers?" It also claims that the problem for discrete-event systems is likely undecidable in general, at the same time mentions particular cases (CSMA/CD, IEEE 802.11), in which formulation and analysis of simple control laws and necessary conditions helped to achieve a solution. We claim that finding the right information is the key to the both problems - decentralised control and its assessment. However in the scenarios, in which the coordination and its assessment need to be achieved between cognitive elements that coexist in different realms (different RATs) the problem gets even more difficult. To discover the essential in control theoretical sense information we look further at the co-design of subjects and objects.

Recall the dependencies within a single policy domain; subject *S* is configured with the positive obligation policy ($O+$) to conditionally request an action a on the target object *T*, i.e. *T(a)*, the object *T* is configured to conditionally authorise this request. We now concentrate on policy conditions; on the subject side the condition triggers the obligation, i.e. the request; on the object side the condition might reject otherwise permitted request. In the majority of practical cases the obligation is triggered by an event (E_O) external to the subject; we introduce another role (Monitor) to represent an entity that generates the event that triggers subject's obligation. On the object side the conditions of otherwise positive authorisation can be those defined by safety, state dependencies, conflict resolution, platform- or device-dependencies, etc. Again, we introduce the Monitor role to be responsible for generating the condition, or without loss of generality just an event (E_A).

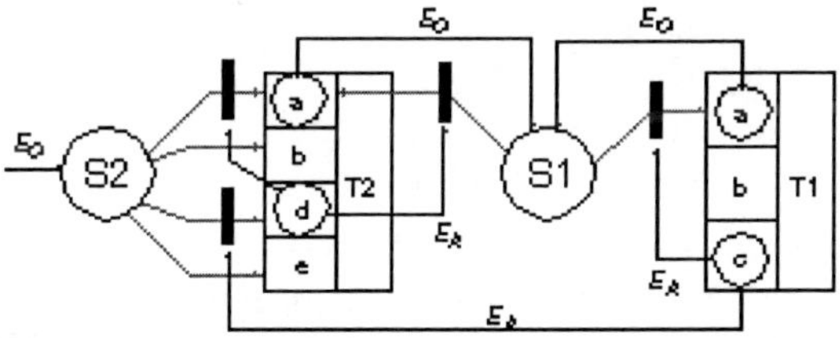

Figure 16. Extending the policy domain with the Monitor role.

The policy domain - a logical collection of subjects and objects that are configured with common policies - does not by definition include the Monitor role. In the design of conventional policy-based systems both types of events were assumed to be generated by the middleware, thus the policy design was limited by the two roles - subjects and objects. We shall now demonstrate how to extend the policy domain concept to include the Monitor role, and how the extended policy domain helps to achieve policy-controllability. Figure 16. shows the dependencies within the extended policy domain: subject S2 is triggered by obligation notification events that are external to the considered example; subject S1 is triggered by the events generated by the invocations of method a of target objects T1 or T2, requests from S1 are accepted conditionally based on authorisation notification events generated by invocations of method T1.c or T2.d respectively. Invocations by S2 of T2.b and T2.e are always authorised, while invocations of

T2.b are controlled by the events generated by the invocations of T2.d; invocations of T2.d by S2 are controlled by the events generated by the invocations of T1.c. The example has no particular meaning besides demonstrating the power of the policy notation that is intentionally borrowing from Petri nets.

In particular, the notation is useful in that it is explicit about all but one dimensions of the event definition; as a consequence the use of this notation will be free from hidden features of RMI semantic. Following the [[27]] the event can be defined a four-tuple: Event = <A, B, C, D>, where A stands for Action, which invocation generates the event; B stands for Box (target object, network element), in which the Action was invoked; C is a set of Conditions, which are considered to be True for the Duration of time D, after the invocation of A on B. All but time attributes of the event can be captured by the proposed notation, however - like in Petri nets - timing aspect can be captured by some sort of reachability (behaviour) tree as explained in the section devoted to the assessment of CRS.

Thus, in the extended policy domain not only subject and object roles are configured with respective policies but the monitor role as well is configured explicitly with the notification policy. Obviously, the notification policy can be classified as positive obligation to notify, thus a set of configured notification policies for a particular instance of a monitor defines its monitoring behaviour.

As it is clear from the previous discussion the instances of the role monitor that notify subjects and objects belong to essentially different realms. Subjects need to be notified on changes in the external environment as a whole (ecosystem, multi-RAT environment), to which changes the subjects need to react (adapt), while objects need to be notified on mainly internal events, derived from the states of resources of a single RAT. In the most generic settings the target objects in our framework represent all types of resources that actually do some useful work; equally generic these resources can be assumed as shared between multiple requests; thus without loss of generality a role object has a property of a multiplexer, and its configured authorisation policy acts as an admission control to a multiplexed resource. At this time we need to look at the extended policy domain from the viewpoint of control theory. From the control theoretical perspective all three roles - subjects, objects, and monitors - are controllers, their pair-wise communications (request, action, event) all contribute to the common control purpose but very differently. These differences are summarised as follows.

Obligation: O+ requests a behaviour change, including the behaviour of sensors (i.e. Monitors) as well; usually such changes in the behaviour are termed adaptations; O- shall make subject to refrain from requesting behaviour change, the purpose of the refrain can be either desired stability and/or robustness of behaviour.

Authorisation: A+ leads to the multiplexing of a request to those already admitted, while A- basically implements the multiplexer fairness - a new request is admitted only if it does not harm the processing of admitted on-going requests. Note that this interpretation of authorisation is networking specific.

Subject notification: a monitor configured with NS+ policy must generate the E_O event in case the configured change in the monitored context is detected (for example, spectrum sensing can detect a frequency band usage opportunity); a monitor configured with NS- shall refrain from generating an event in case of uncertainty or conflict.

Object notification: a monitor configured with NT+ policy generates the E_A event when resource usage change is detected, thus signalling the opportunity of new requests to be admitted; a monitor configured with NT- shall refrain from generating an event in case of uncertainty or conflict.

Following the above semantics we consider an extended policy domain as a distributed controller system with the three types of information exchange (within the system and with the environment). First, it is the behaviour itself - externally observable set of actions; second, it is the sensing behaviour, and, third, it is the exchange of events within the system and with the environment. However, to be able to enforce and to maintain the control purpose we need yet another extension of the policy domain - we need to include the cognition, which will facilitate the required coordination between different subjects. This extension is achieved by the diversification of the Monitor role and the use of composition principle.

The recent work [[28]] among other works reports a success in enforcing collaboration between neighbour agents by introducing so called remote policies into a local agent's ruleset; as the main conclusion this work generalises that "multi-policy scenarios outperform single-policy scenarios". Indeed, our distributed controller system is created just with the aim to facilitate these multi-policy scenarios. The first extension of the policy domain was made above by introducing a new role - Monitor of two types with the subsequent notification policies of two modalities each; the second extension of the policy domain is the diversification of the monitoring behaviour. Let us distinguish between objective and subjective monitoring as below.

Objective monitoring is the behaviour of a CRS defined by its obligation, notification and authorisation policies, that is realised in adaptive sensing and subsequent usage of radio resources as defined by its policies. The objective monitoring can be seen on Figure 17 as a sequence of compositions of the role subject of that CRS with multiple instances of the role Monitor. The role Monitor can be assigned to RAT sensors, to platform monitors (CPU usage, system clock, etc.), to generic context sensors (e.g. GPS-based positioning), etc., abstracted as a Resource.

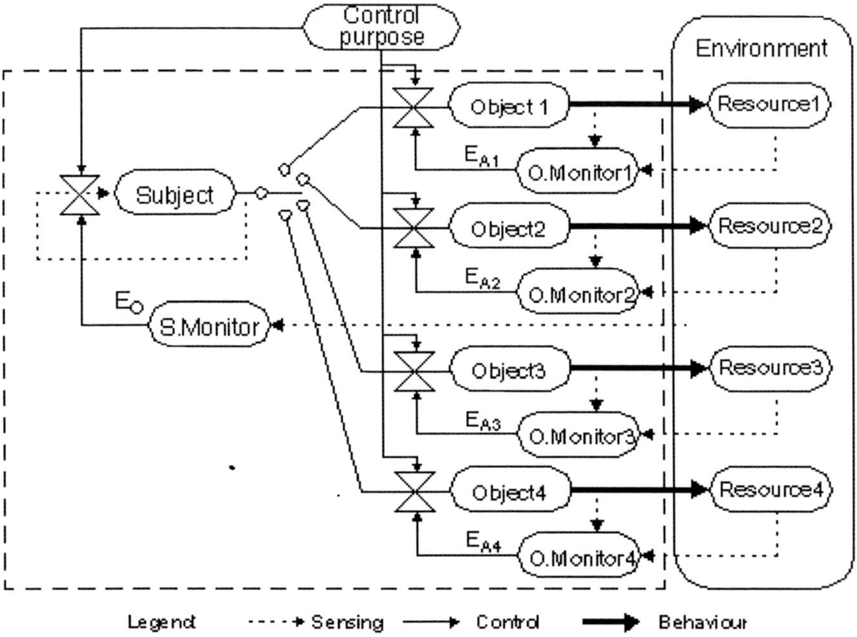

Figure 17. Objective monitoring as a composition.

In other words, objective monitoring is a process helping CRS to adapt its own behaviour in reaction to sensed changes in the environment.

Subjective monitoring is the behaviour of a CRS that decides on the changes in objective monitoring; it is based on cognition, it is responsible for recognition, comparison, attention, etc. all cognitive actions generally termed perception. It is the subjective monitoring process that infers the policies of other CRS's, and by aligning its own policies eventually participates in the creation and maintenance of the etiquette - coordinated and informed mutual adaptations of multiple CRS sharing the same time, space, and spectrum resources. Indeed, a set of CRS's federated by a common etiquette is a multi policy scenario. Note that subjective monitoring is required to be cognition-based even in the presence of a facilitator like Cognitive Pilot Channel (CPC) being standardised by the IEEE P1900.4 since

2007[17], simply because any recommendation heard by a CRS on a CPC might be in conflict with CRS's own policies, such as those defined by a device manufacturer, user, home and visiting network operators, home and visiting regulators, etc. Let us look at a couple of use cases that are common place in the body of current work on self-optimization in heterogeneous wireless networks, and demonstrate the essence of subjective monitoring based on these use cases.

The first use case is "Load Balancing"; a Multi-RAT base station (MBS) is assumed to be able to re-assign its own resources between RATs. For this to happen it is required that the policy database at MBS keeps the measured load per RAT and relates these loads to load balancing thresholds. Assumed is that the above policies trigger the reconfiguration process between RATs at the MBS.

The second use case is "Energy Saving"; at low traffic in single or multi-RAT environment network operator would be able to automate the process of switching redundant at the moment bases stations on and off. It is required that network operator uses its policy database to relate the measured load to the load predictions per cell and create the needed on/off policies. Assumed are: i/. the process that controls on/off of the cells; ii/. that different network operators can cooperate; iii/. that QoS is met always; iv/. the "Load Balancing" use case can coexist with it.

Normally, designers consider a set of use cases that is complete in a sense that all wanted behaviour aspects of a system under design are covered. This set can be seen as important input for the requirements capture, and further design steps that are detailing the invisible at the use case level internal mechanisms of the system. Per convention, a use case describes behaviours that are triggered by some events, and thus are described as system reactions on stimuli. However, if a system has learning capabilities within its decision process the above relation of triggers to behaviours may become opaque, or at least not easy to capture. We show now by the interaction of just two use cases above, how subjective monitoring helps to avoid policy conflicts and unwanted policy interactions during policy multiplexing.

What happens in a network realising the above use cases concurrently? Imagine a user terminal that is triggered for a handover (mobility event) and surrounding MBS's are being triggered for a load balancing at the same time; it may happen that the terminal's handover policy will be in a conflict with the policies of other network elements. Another obvious danger is when a cell that is the target of the load balancing enforces the switch-off policy. The problems of policy conflict detection and conflict resolution are known to be hard; multiple

[17] http://standards.ieee.org/announcements/PR_P1900.4_New.html

approaches to these problems reported in the literature address these problems off-line (for example, applying model checking techniques). In complex dynamic systems one needs methods for the run-time detection and resolution of policy conflicts.

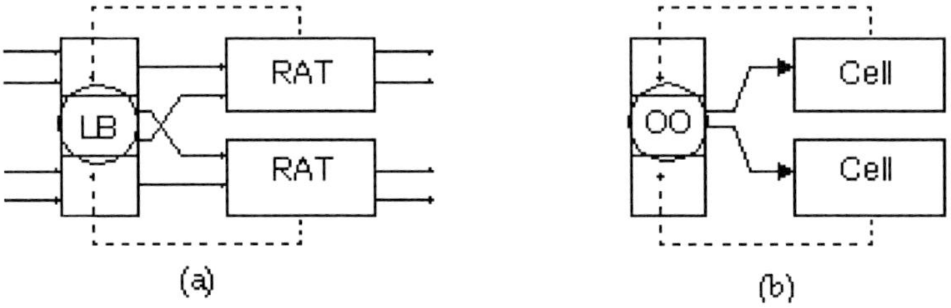

Figure 18. Use cases: Load Balancing (a), and Energy Saving (b).

Figure 18 uses the notation to describe graphically the two use cases. In the load balancing case the obligation policy element (subject) is labelled LB, which stands for Load Balancer; here the policy element acts effectively as a payload switch between the two RATs. The model on Figure 18.a shows that sensing data is being related to the thresholds inside the policy element; the reconfiguration policies computed by the policy element and disseminated to the respective mobile terminals are not shown. In the Energy Saving use case the obligation policy element acts effectively as On-Off switch for respective cells, hence it is labelled OO; the payload data flows are abstracted on Figure 18.b however the control connectors, by which a policy element switches cells on and off are explicitly shown; dashed lines represent notification data, i.e. cell/RAT load.

In a multi-RAT environment the two use cases can coexist, and moreover can be triggered for the concurrent execution. It also may happen, that a cell, belonging to a under utilised RAT will be requested to switch off by the OO policy element. Obviously, if it happens that in the same area the under utilised RAT coexist with the one that is over utilised the LB policy element can decide to switch part of the payload traffic to the cell that is just requested to switch off. The concurrent and not coordinated decisions of LB and OO can easily create a racing condition. A careful engineering of the two use cases will avoid the above conflict provided that the engineering of both policy elements is in the same pair of hands, which cannot be always expected in reality.

In the case of the concurrent invocation of the two use cases, we model (as on Figure 19) the conflict by placing corresponding cells inside the respective RATs. Both policy elements receive policy data (the upper RAT / Cell is under utilised) from sensing of the load within the upper RAT and respective cell, which data triggers the two decisions - to switch part of the traffic from lower RAT to the upper RAT by the LB and to switch off the under utilised Cell by the OO. Not only the notation allows to make potential conflicts clearly visible but also it suggests a way out of the conflict.

The logic of the conflict resolution is simple: if LB (and vice versa the OO) would know about the decision that they are going to enforce, and if these decisions are mutually exclusive (in our case that would mean that the Cell being switched off must be exactly the cell that would be serving the balanced traffic) then one or both policies could be adjusted (multiplexed in a conflict free manner) to avoid the conflict. This exchange of these "would be enforced policies" made explicitly visible by the respective arrows on Figure 19.

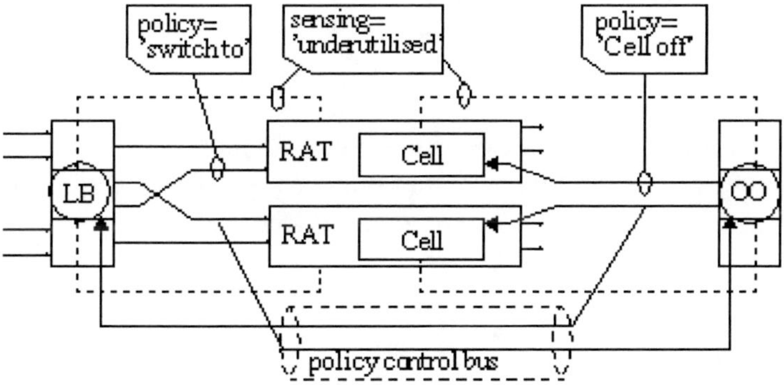

Figure 19. Policy conflict detection and resolution example.

The conflict resolution approach demonstrated in this example can be seen as an ad hoc one, however (in line with the promise theory adopted by the IRTF DMRG) it appears very generic. One could think of a network wide "policy control bus", to which multiple and heterogeneous policy computation points would publish their "would be policies" (i.e. policies that are likely to be enforced in a near future) and would listen to comments from other subscribers and make respective policy adjustments. Such an approach is in a striking similarity to the flavour of multicast communication known as "claim-collide" and is standardised by the IETF MALLOC for the distributed and automatic allocation of multicast addresses: domains claim chunks of addresses and if no collision is reported back the claim is considered as a safe allocation.

The above described policy control bus, or CPC as its particular flavour are indeed good facilitators of spectrum etiquette, however as it was mentioned above the inference capabilities are still required. The exchange of intended policies has certain latency, which cannot be negligible in all cases, the next section shall show a radical example of cognition assisted decision making in absence of any explicit information exchange.

Eco Red: Example of Self-learning Policy

In situations that require from a CRS quick decisions it is often not possible to run the full-scale decision process that through sensing, decision making, and adaptation would implement the optimal behaviour. Is it still possible to use policy as the main control mechanism in these situations? Do we know useful cognition patterns that can improve the utility of policy through its own learning? Typical situations of that type are found at the level of datagram handling within various network elements; not surprisingly the first mentioning of self-learning policy is found in ground-breaking work of P. Baran . Following this early pattern of self-learning (inferring the correct routing table entry from observations of transit traffic and simple computations of certain counters) this section attempts to build a framework for other than routing policies. As a baseline example we use active buffer management scheme, known as random early detection with our eco extension sketched earlier.

Entropy at Work

Recall that our main interest is multiplexing, more precise - fair multiplexing. Fairness - conformity with rules or standards - is an important attribute of any service, therefore the fairness of multiplexing in relation to policy should always be evaluated. Many traffic handling mechanisms found in the current Internet (for example, buffer management in a router, base station, relay point, etc.) are using policies implicitly - not exposed to end points; traffic multiplexing makes it hard, if not impossible to enforce explicit policies other than end-to-end. Over time the end-to-end fairness was substituted by the notion of TCP friendliness [[29]]. used to evaluate and control non-TCP applications. As before we consider the smallest possible ecosystem - a single buffer in a router populated with datagrams of different connections, and demonstrate now how the ecosystem approach helps to design truly fair and autonomic buffer management scheme; moreover we show how this scheme can learn and enforce various policies.

Congestion avoidance is that part of TCP behaviour, which is responsible for the notion of TCP fairness; for a best-effort treatment it is basically the fairness of a FIFO discipline of a buffer. Consider first a single buffer and assume that the buffer is shared by a number of active connections; in case of congestion (seen as buffer overflow) all active connections will experience similar packet loss, congestion avoidance mechanism of TCP will force all connections to shift to the slow start (i.e. to halve the sending rate) - notably, this will happen nearly synchronously for all affected connections. Assuming now that all connections continue and the routing does not change and that new connections did not arrive the previously over flown buffer and the congested link will become underutilised. Eventually, the slow start will, under all the assumptions above nearly synchronously bring again the node to the congested state. One popular in-network solution for the problem is known as RED - random early detection [[6]]; it removes the synchronisation between slow downs of active connections by dropping random datagrams before the congestion occurs. For this RED monitors the statistical properties of a buffer. As [[7]] states the simple "RED makes Quality of Service differentiation impossible", however, perhaps more fundamental is that RED breaks the fairness that is somehow preserved in synchronous slow-down. In the text surrounding Figure 1 we discuss that straightforward enhancements of RED with application level policies are not possible.

Bearing in mind that RED was proposed as a local queue management algorithm that makes decisions only on local observations the above application policy would be foreign to it; additionally, since each application is allowed to set the highest criticality for its every connection the above policy enhancement is useless. Instead, RED could be modified in a way that considers a buffer as a micro ecosystem populated by datagrams of active connections; the connections should be viewed as yet another ecosystem, and so on up to the ecosystem of interconnected networks, which in the Internet slang are Autonomous Systems (AS). An AS is an arbitrary network defined by its single and uniform routing policy. The transit traffic handled by an AS is convenient to represent by a traffic matrix - the major input to AS's traffic engineering (TE) process. The detailed TE discussion can be found elsewhere, here it is important to understand the relation of ecosystem policies at micro and macro levels, and to use this relation for the enhancements of the former. As before the desired RED enhancement problem can be solved if formulated as a control theoretical one.

The above control problem can be used for RED enhancements in somewhat inverted form. Let us assume that the above tuples of controllers are TCP congestion avoidance mechanisms located at endpoints of each connection; the information to be exchanged between controllers is specified by the TCP, and is transmitted through each buffer (and associated RED instance) on the path from source to destination. The major question now translates into "What information should RED collect to make fair drop decisions?" Since a micro-ecosystem (buffer controlled by RED) can be seen as a multiplexer then the RED can be seen as a capacity admission control of that multiplexer (interpreting dropped datagrams as if rejected on their admission attempts does not change the end result). A multiplexer is known to be fair if admission of any new input does not change the service rate received by all admitted requests. For a buffer the service rate for admitted requests is connection's effective bandwidth on the associated link. One cannot define that datagrams in the buffer are admitted requests since RED can drop them before the service will be received, however one could construct a control mechanism based on buffer sampling that will assure the fairness.

Let us assume that for each active connection the associated traffic process served by the micro ecosystem is estimated by an empirical rate (r) and Packet Loss Ratio (PLR) that is a real value within [0,1]. Figure 2.a shows how the micro ecosystem is modelled as a simple multiplexer with L incoming traffic processes multiplexed on a single line with the permanent [link] service rate s; a multiplexer has finite buffer of size b, the instant buffer occupancy is denoted by q. Assuming that target QoS parameters, such as PLR denoted by c are known, the expression for the link bandwidth $B(q,c)$ requirement is

$$B(q,c) = min(s \mid PLR(q,s) \leq c)$$

RED considered as a multiplexer needs to relate the above expression for link bandwidth requirement to drop decision provided the condition of pre-congestion is detected.

Such a relation (a safe bound for $f(L,b,s)$, the frequency of packet loss when L lines input a buffer of size b emptied at rate s) is given by (1), a Duffield-Buffet formula (see e.g. [[30]]):

$$f(L,b,s) \leq \exp\left(-L\mu\left(\frac{s}{L}\right)\right)\exp\left(-b\delta\left(\frac{s}{L}\right)\right) \tag{1}$$

The formula was obtained by modelling bursty ATM traffic by the two-state (active and dead for each input line) Markov model with probabilities a and d of switching between the states. The formula is remarkable since it demonstrates that the log-frequency of packet loss versus buffer size is less than a straight line with slope $-b$ and intercept $L\mu$ (see Figure 2.c). The coefficients μ and δ can be obtained numerically from probabilities a and d, they do not depend on b but only on the ratio of s to L, that is the effective bandwidth. Since [[30]] assumes $L = const$, which is hardly the case in a router buffer the usage of (1) for RED drop decisions requires some deliberations, though such usage is indeed valid since PLR does not depend on L as well.

The enhanced RED Figure 2.a communicates its decision to TCP congestion avoidance mechanism (TCP CA) of a connection implicitly by dropping connection's datagram. The drop decision process can be sketched as follows. Though RED has practically only one buffer, it however monitors the buffer occupancy through e.g. standards track IPFIX technology equipped with appropriate packet sampling algorithm. This monitoring collects statistics of active flows in the flow table (Figure 1.b); in the simplest case this statistic is just datagram count over fixed time interval, hence the estimation of connection's rate (r_i at Figure 2.b) is its packet count over this time. An active connection contributes to overall buffer occupancy with the amount q_i of its datagrams that are in the buffer at the decision instance, say t. Let L_t denote the instant number of active connections at the decision time, then the ratio of s to L_t is the fair effective bandwidth for an active connection, and $r_i /(s/L_t)$ defines the relation of connection's rate to the fair effective bandwidth. Consequently, RED should drop a datagram of the connection with the highest value of the ratio of q_i - their share in the current buffer occupancy, to $L_t \cdot r_i$, where r_i is the rate estimation taken periodically.

Why this drop decision is fair[18]? The target PLR c of a connection is not known to RED however its actual value for all multiplexed flows can be readily estimated by computing the asymptotic log probability's slope value $L_t \cdot \mu = (c'' - c')/(q'' - q')$ (Figure 2.a shows current buffer occupancy q and the two its samples that define the straight line on Figure 2.c). Assume that such PLR estimation is done periodically, i.e. synchronously with the periodical rate estimation, meaning that the value of r_i is that from the previous estimation interval, while both q_i and L_t are fresh, they are instant values at the decision moment. The datagram selected for the drop will be either the one from a new connection ($r_i = 0$) - in this case the drop decision is as fair as the capacity admission control is fair; or it will be from a connection that either had the highest rate in the previous observation interval, or has the highest rate currently (as indicated by the high value of q_i), or both.

---

[18] The next sub-section shows a convincing numerical example demonstrating this clearly

Even a single buffer interpreted as an ecosystem shows that its behaviour can be controlled in the way that supports sustainable growth of the whole ecosystem; in this particular case meaning preferential treatment of long-lived connections (in other words, supporting a sustainable datagram handling workflow) constrained by the adherence to the effective bandwidth share that is fair for the existing population of sustainable workflows, and sensitivity to burstiness increase of a connection somehow captured by relating past rate to the current datagram count.

From the higher level viewpoint the system consisting of a number of TCP CAs multiplexed over the enhanced RED exhibits complex behavioural dynamics that is defined by implicit model-driven information coupling of distributed controllers contributing to the same control purpose though having completely different viewpoints on the controlled reality. The information coupling is achieved through the fairness based on a model - a linear safe bound of PLR, a critical performance indicator.

These qualitative considerations as well as the underlying mathematics (expression (1) in particular) are clearly pointing at entropy of a system of a set of TCP CA and a buffer; the higher the entropy of the system the more needed is the drop decision. The drop decision needs however to be fair with regard to the entire ecosystem; intuitively, it appears that the fair decision will be the one that takes into account the root cause of the changes in the system's entropy. How a decision process being part of an ecosystem can learn its own entropy?

Learning Own Entropy

Earlier, we noted that a policy coupling between subjects and objects is somewhat analogous to the chemical bond, and as such can be seen as having certain energy in that bond, which can be measured objectively and used in decision making. Thermodynamics defines entropy as randomness - a quantity representing the amount of energy in a system that is [no longer, or not yet] available for doing mechanical work, while information (communication) theory defines the entropy as a numerical measure of the uncertainty of an outcome. Without diving into a debate about the relation between the two definitions let us just apply the Shannon's formula to our ecosystem, and see how with this formula our EcoRED can learn its own entropy. In doing so, we follow the experiment described in [[31]] that proves that Shannon's entropy is an excellent predictor in avian visual cognition.

Shannon's entropy definition is given by the expression (2), which we interpret (borrowing from [[31]]) as follows. A closed[19] system - a buffer with RED - is considered as a categorical variable denoted *A*, that is a metric of a characteristic that has been measured on the nominal scale. Note, that despite the fact that nominal scales are used to measure non-quantitative things while we plan to measure buffer occupancy (very much quantitative things), there should be no confusion. We shall use numerals to infer the entropy denoted *H(A),* which is essentially categorical[20].

$$H(A) = -\sum_{a \in A} p_a \cdot \log p_a \tag{2}$$

In the expression (2) a is a category of A, and p_a is the proportion of observed values within this category. We interpret an active flow, which has at least one datagram in the buffer as a category a, meaning that p_a is a measure of diversity. If the buffer is shared by more than one flow, the p_a must be computed as the number of datagrams of one active flow in the buffer divided by the total count of datagrams present in the buffer at the time of computation; otherwise p_a =1. An active flow is conventionally defined as a five-tuple < Source IP Address, Destination IP Address, Source Port Number, Destination Port Number, Protocol Number>. We treat a simple datagram count as a category because we want to infer unknown pattern of each active flow, which pattern is observed at a buffer as a count however in the RED decision process we need to use this pattern for effective bandwidth estimation.

To have a numerical baseline example let us consider a router buffer that can handle on average 256 datagrams assuming certain normalised datagram size. Let us also assume that the line[21] speed is such that the buffer can be fairly shared by the 16 active sessions, each session having such sending rate that at the moments of periodic buffer samples there are at most 16 datagrams of each active flow. It will become obvious later that these assumptions (or should we better term them policies?) are really redundant, however at this time they serve as important mind setting.

The important mind setting is the differentiation within a buffer between Self and Non-self. For each active flow all datagrams pertaining to that flow are said to be self, while a datagram pertaining to any other flow is said to be non-self. When

---

[19] An open system in this case shall include all TCP CA mechanisms for each active flow

[20] At least in the thermodynamics sense

[21] To which line our EcoRED serves as a fair multiplexer

our buffer is shared by a single flow then the flow can be sent at the rate equivalent to buffer occupancy up to 256 datagrams; regardless the amount of datagrams a=1, and since we use categorical variable p_a =1, and log(p_a)=0. In the single-flow case EcoRED decision described in the previous sub-section does not require any cognition: when early congestion threshold is met by the buffer occupancy any dropped datagram will be the self one. Another trivial drop decision corresponds to the case when a buffer is shared by up to 16 equi-rate flows, meaning that up to 16 flows might have concurrently up to 16 datagrams in the buffer, and all datagram counts are equal. The decision in this case can be done on random. Obviously, in case all datagram counts equal 16 the $H(A)$=4, which is maximal value in our case, all other interesting cases involving cognition are within $H(A)$={0..4}.

The cases for cognition are those, when a drop decision is not obvious. Let us see whether we can use the entropy as a robust indicator of a correct (fair) drop decision. For this we need a [nominal] scale for systematic exploration of the entropy behaviour within the range defined above. Fortunately, since the value of entropy is not[22] sensitive to permutations within the buffer of datagrams belonging to different flows, such scale is feasible.

Since it is useless to explore all possible permutations of datagrams for each level of buffer occupancy, what can be another useful technique to study the entropy? Obviously useful is the above Self/Non-self (S/N) differentiation: one could study the behaviour of entropy function in different S/N cases. Those cases allow easy enumeration following the pattern NSS/NNN, where NS stands for the datagram count for Self and NN - is the datagram count for Non-self. For example, 5S/12N yields the value $H(A)$={3.40..0.87} for the distributions of datagram counts (DC) per virtual buffer (VB) shown in Table 2, assuming that a virtual buffer is assigned per flow.

Table 2. Entropy values for 5S/12N

VB	1	2	3	4	5	6	7	8	9	10	11	12	13	14	15	16	H(A)
DC	5	1	1	1	1	1	1	1	1	1	1	1	1	0	0	0	3.40
	5	1	1	1	1	1	1	1	1	1	1	2	0	0	0	0	3.28
	5	1	1	1	1	1	1	1	1	1	3	0	0	0	0	0	3.12
	...																
	5	12	0	0	0	0	0	0	0	0	0	0	0	0	0	0	0.87

[22] Which is obvious but adds an argument against the treatment of entropy as a disorder

Thus, we conclude that entropy interpretation in the decision process must be somewhat situative, meaning that low entropy values bear much more information for decision making when fewer flows are present in the buffer than higher entropy values with many flows. Indeed, the last row of Table 2 indicates that if the congestion occurs in the next sampling interval with still only two flows active then the flow occupying the second VB is a clear candidate for a drop.

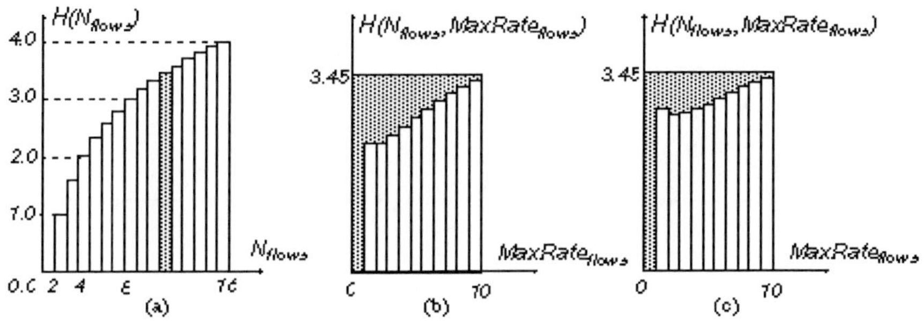

Figure 20. Entropy behaviour as a function of: number of active flows (a), number of flows increasing the maximal rate to 16 with N_{flows}=11 (b), number of flows increasing the maximal rate to 8 with N_{flows} =11 (b).

Continuing this systematic study of entropy behaviour, we show the plot (Figure 20.a) of maximal values of entropy - the entropy here is a function of N_{flows} - the number of active flows, and it reaches the maximum value when all active flows are equi-rate ones. Let us fix the number of flows, for example N_{flows} = 11, which gives $H(A)_{max}$=3.45 for the case of equi-rate flows, for example r=1. Let us see how the entropy changes when the rate of the Self flow (which is always assumed to occupy the first VB, but the order is arbitrary, of course) remains r=1, while growing but constrained by the N_{flows} = 11 number of Non-self flows increases their rate.

If Non-self flows increase one by one their rate from r=1 to r=16 the entropy will first drop from 3.45 to 2.23 and then gradually increase up to 3.35 (Figure 20.b); if Non-self flows increase their rate from r =1 to r=8 the entropy will first drop from 3.45 to 2.83 then down to 2.72 and then gradually increase up to 3.37 (Figure 20.c).

Since any combination of the same number (NSS/NNN) of non equi-rate flows yields lower entropy then the same number of equi-rate flows it appears reasonable to show these lower values as being "inside" the equi-rate entropy value. This is implemented on the figures by expanding the shaded bar corresponding to equi-rate entropy and by putting all studied non equi-rate values inside that expanded box. The numerical experiments similar to those presented on Figure 20 for $N_{flows} = 11$ when done for $N_{flows} < 11$ show that observed entropy drop and subsequent growth can be fairly approximated by the linear saw-tooth pattern shown on Figure 21.

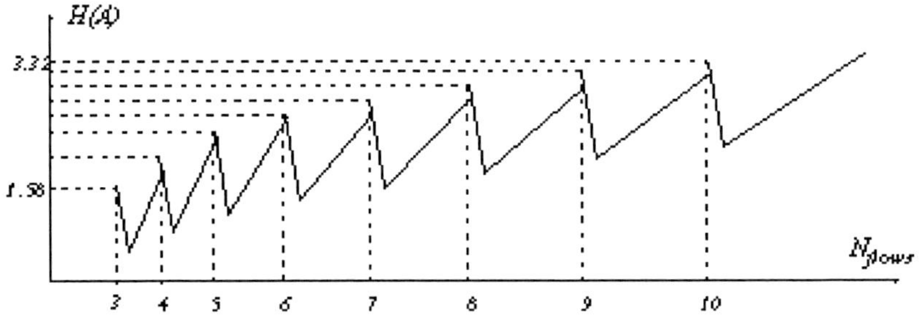

Figure 21. Entropy linear pattern.

This pattern is in the striking similarity to the saw-tooth pattern of TCP behaviour that is induced by its congestion avoidance mechanism shown on Figure 1. It appears reasonable to conjecture that entropy-based self-coordination of flows (between TCP CA and EcoRED) leading to fair drop decisions is indeed possible.

To conclude this numerical example, let us investigate the nature of non-linearity that appears evident when the buffer is shared by more than eleven flows.

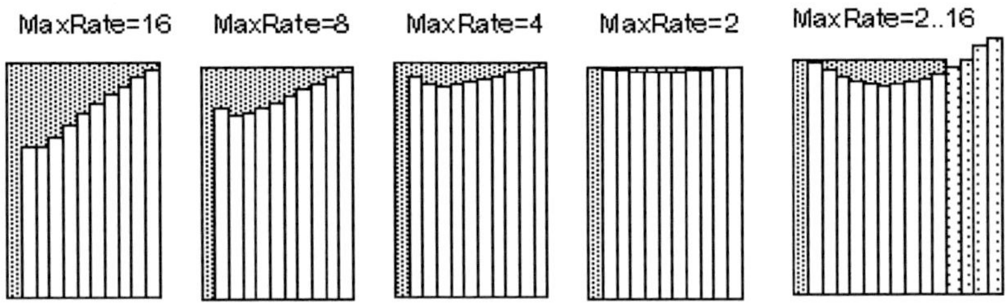

Figure 22. Entropy non-linearity.

Figure 22 shows entropy non-linear behaviour, when starting with 11 equi-rate flows of rate r we increase the rate of Non-self flows to a maximum level that is, left to right $16{\cdot}r$, $8{\cdot}r$, $4{\cdot}r$, $2{\cdot}r$. The rightmost diagram on Figure 22 is somewhat different, it shows the entropy behaviour for the case of maximal diversity, it also violates the limitation $N_{flows} = 11$ and demonstrates nearly symmetrical growth of entropy values. The maximal diversity case is the one that starts with 11 equi-rate flows and increments the increase of each `next to burst' flow, thus the rates on the last diagram comprise the sequence $\{2 \cdot r, 3 \cdot r, ..., 16 \cdot r\}$.

As the main finding of this section let us state the importance of learning of own entropy behaviour as a function of major parameters that describe the situation, in which a decision needs to be made. Entropy appears to be a good predictor, yet it needs to be related to the policies that represent the control purpose. In generic cases (like fairness policy studied here) some generic solutions (like entropy saw tooth behaviour) are possible, in more complex cases a CRS must be self-aware by knowing how to assess the situation.

Policy Agility and Control Loops

Much discussed policy agility property is the challenge of cognitive networking: in particular CRS must not only embed cognition in the radio control loop[s] but also not to embed policies into the radio; the policy agility requires that policies are learned, and if needed forgot, and when needed remembered. We have formulated this challenge rigorously as a control theoretical problem, which in turn allows us to describe formally the proposed assessment framework in the next section. This short section aims to describe who needs the assessment, and what are the most important requirements for this.

We state that the CRS empowerment required from the cognition is not the intelligence *per se* (assessing the amount of knowledge is not required), however the ability to solve domain-specific problems (assessing the know-how is required). Until now we have concentrated on the design of CRS, however after the design is done, do we know how to assess the properties of a self-managed system? Additionally to performance evaluation and testing to guarantee for such systems consistency, safety, purposefulness, security, and efficiency of their operation engineers require assessment - systematic evaluation [of non-conventional to previously engineered systems] properties such as cognition, learning, aware sensing, and perception.

According to the studies reported in [[32]] practically deployable CRS techniques should address the two sets of requirements - technical (including radio resource management, reconfiguration and efficiency) and non-technical (regulatory and business aspects). At the same time, thus converged [wireless] service infrastructures shall exhibit a novel property - they shall facilitate the secondary market [[34]] this future secondary market is compared to existing energy secondary market, where the exchange demand exists on hourly basis, with the conclusion that "licensed spectrum commodity secondary market might have a near-real-time exchange demand". It is clear that conventional control and management techniques cannot be applied in these likely future scenarios; it is also commonly agreed (after the pioneering XG initiative of DARPA [[33]] that needed control mechanisms should be devised within the policy domain.

DARPA XG program was perhaps the first large-scale effort to address the issue of policy-based management for CRS, and perhaps the first to realise that the required unbundling of functionality and control can be realised if [network element] behaviour is considered as a resource. The [[35]] argues that the novel architecture for FSM must meet the following requirement "A core set of behaviours must be identified in such a manner that a viable architecture where only the core set needs to be considered for regulatory approval is possible" and defines the following top-level abstract behaviours (AB): sensing, identification, dissemination, allocation and use of opportunities. These abstract behaviours are considered to be similar to a secure kernel in conventional computing: kernel behaviour is completely predictable, while the outer behaviour can be any.

The [[33]] outlines important requirements for CRS policies, such as policy consistency, adaptation of radios to policies, openness for new capabilities, and the stressed earlier separation of policy and behaviours. The latter requirement is essential for the understanding of the innovation of our work, hence we quote [[33]]: "It is a goal of the XG program to separate the policies that govern an XG system from the behaviours of the system. This will ease accreditation of XG systems by enabling regulators to accredit radios based on the ability to interpret policies correctly to obtain desired behaviour, not (as today) by conformance of the implementation to specific pre-determined policies. These behaviours are discussed in the XG Abstract Behaviours RFC [XGAB23]. In the absence of such separation, accreditation would need to be performed every time a policy was changed even in a minor way."

There are several issues with the above requirement that sketches the accreditation (which is obviously relevant to the assessment we present later); we try to address these issues in our assessment approach. Firstly and fundamentally, it is not clear how a policy, when understood (after M. Sloman) as a rule defining a choice in the behaviour of a system can be separated from a behaviour that hosts choices; our take on this is to consider concurrently temporal and spatial aspects of the above separation. Secondly, the required ability to interpret policies correctly is proposed in [[33]] to be verified by the comparison of resulting behaviour with the desired behaviour, meaning practically that policy set as such is outside the accreditation. In our approach not only resulting behaviour has to be assessed but the set of given policies as well; we put forward the compliance of the both to the situation, which is defined above and used below. Lastly, since XG is silent about abstract behaviours it appears that the missing notion of situation is critical. Our assessment framework considers permanent verification of a behaviour based on the observation of the exhibited behaviour, of given policies and of the situation, the latter includes the current System under Assessment (SuA). In our approach (inspired by the control theory and theoretical computer science) the given (external to the SuA) policies are considered as peer entities to that [functional] behaviours that host choices controlled by policies.We consider the SuA's resulting behaviour as the behavioural composition of the above two types of peers. In doing so we agree with the statement in [[36]] that in the behavioural composition "the constraints can be represented like any another component behaviour", and further extend the approach to consider any policy as a [functional] behaviour constraint.

In considering spatial and temporal separation of policies from behaviours we also follow the trend known in wireless communication as spatial multiplexing; for example, talking about the need to exploit the waveform diversity the [[37]] states that "Link" is no longer just an input to control mechanisms (such as routing protocols), but also a parameter that can and needs to be controlled, which leads to a challenging multi-dimensional topology control problem, and opens an opportunity of distributed real-time optimal algorithms for "custom made" topologies." Our behavioural composition of policy behaviours and functional behaviours is formulated as the above topology control problem but in the control-theoretical sense.

CRS Assessment Framework

Cognition and Assessment

Our distributed controller system - an interpretation of the extended policy domain - is essentially a composite system; thus its assessment can proceed following the "specifically objective" frame of reference introduced by G. Rasch. However, while this is sufficient it is not the necessary condition to claim that exactly this composition is the correct one for the assessment of cognitive radio systems. This brings us to the issue of cognition in cognitive radio systems; without trying to define what cognition is, we instead try to define what is the purpose of cognition in cognitive radio, meaning for the assessment that if the purpose of cognition is met the assessment of underlying cognition can be high, and low otherwise.

It is important to compare now the assessment of cognitive radio networks with their performance evaluation. Not surprisingly, the performance evaluation chapter of [[4]] asks somewhat different but very relevant and useful questions. They are: What happens when we deploy cognitive radios? Will they perform as we expect when faced with a realistic environment? And exactly what is realistic environment?

To answer these questions the quoted authors take slightly different view on cognitive radio from co-authors of other chapters - instead of using cognition cycle to explain the operation the chapter considers the cognition cycle as a limited model of the outside world ("the environment, in which the cognitive radios are observing, learning and reacting"). The chapter says "A more realistic view of the operation of cognitive radios would depict an outside world whose state is jointly determined by the adaptations of several cognitive radios... cognitive radios react to an outside world determined by both "dumb" and other cognitive radios" - the former is much more difficult to understand than the latter. Later the chapter states "abstractly, this interaction can be viewed as a recursive interaction decision process in which each adaptation can spawn an infinite sequence of adaptations" [[4]].

This type of reasoning when applied for the assessment allows us to map this recursive interaction decision process onto our extended policy domain model - our distributed controller system is not seeking point correctness but rather process correctness [[38]] of the behaviour. The coexistence issue of dumb and cognitive radios we broaden by the assumption that cognitive radios might be heterogeneous; this assumption is captured by allowing uncertainty as an outcome of the monitoring behaviour. In full accordance with [[4]] our model of the external world used by a single cognitive radio for its adaptations is built and maintained by subjective and objective monitoring of adaptations - own, recorded as a history of behaviours, and of others - recorded as the sensing of multiplexed resources.

We define ideal cognition as the one that memorises all possible situations; assuming unlimited resources for the decision making the ideal cognition always helps to select the best match behaviour based on the past experience, which includes sufficient training. Assessment applied to the ideal cognition should yield the highest value. Implicitly our definition of ideal cognition assumes also the ideal decision process, which in the above ideal settings can be considered as the problem of pattern recognition without noise - fairly easy task for any search algorithm. In reality we must assume that cognition (or the training of it) is not exhaustive, that decision process must act in uncertainty and within constraints, and that sensing is being performed in noisy environments. All these factors contribute to the uncertainty and can be captured by the confidence level, which is our characterisation of the situation presented to the decision process. At the same time the decision process itself can be characterised by its ability to produce more or less correct decisions, where the degree of correctness will be of course judged on the delivered performance. Since Rasch model was already spelled as the right framework for the assessment of composite systems let us briefly describe it.

Specific Objectivity Framework

The term "specific objectivity" and the corresponding method were developed in 1960's by Georg Rasch, a statistician from Copenhagen, while he was studying reading disabilities exhibited by various groups of humans. Rasch Model provides a framework for comparison of test outcomes (*h*) produced by different humans, namely as shown by expression (3),

$$h_{ni} = z_n \cdot E_i \tag{3}$$

where z_n is the ability parameter of the *n*-th human; and E_i is the easiness parameter of the *i*-th test. We use the opportunity of this sub-section to formulate as dense as possible but closely following the logic of [[39]] the specific objectivity framework in its basic form.

The basic or deteminate framework ($\boldsymbol{F}_d$) considers the two sets of elements - the set of objects (***O***) and the set of agents (***A***) - that both appear as factors in the comparison. Any object may enter into a well defined contact with any agent, which contact may yield certain result; all results are said to form the set of results (***R***). Hence the basic framework is given by expression (4). The framework $\boldsymbol{F}_d$ is termed determinate since it is required that each contact determines the result uniquely as in (5), where *r* is a reaction function of both *O* from ***O*** and *A* from ***A***. Since ***O***, ***A*** and ***R*** can be qualitative concepts the reaction function *r* is a single-valued mapping, also shown in (5).

Following (5) any two objects O_1 and O_2 can be compared with regard to their reactions to contacting agent *A*, the values of this comparing function $u(R_1, R_2)$ form a collection ***U***, the values of which can be also qualitative. In reality, the values of ***U*** are statements about the objects O_1 and O_2 as it is shown in (6), which also uses the notation for the comparator v explicitly conditioned on the agent A used for the comparison, which is said to be a local comparison in $\boldsymbol{F}_d$.

$$\boldsymbol{F}_d = \langle \boldsymbol{O}, \boldsymbol{A}, \boldsymbol{R} \rangle \tag{4}$$

$$R = r(O,A) \qquad r = \boldsymbol{O} \times \boldsymbol{A} \rightarrow \boldsymbol{R} \tag{5}$$

$$u(R_1, R_2) = u(r(O_1, A), r(O_2, A)) = v(O_1, O_2 | A) \tag{6}$$

For the global comparison in $\boldsymbol{F}_d$ the comparator v should not depend on the instance of A; if this happens then our pair-wise comparison of objects is said to be specifically objective. The meaning of objective is twofold; first, the comparison of objects O_1 and O_2 does not depend on the instance of A, and, second it does not depend on other objects from $\boldsymbol{O}$. The meaning of specifically denotes that the comparison is specific to the framework $\boldsymbol{F}_d$. As it is demonstrated in [[39]] and in multiple on-line publications the global comparison is indeed possible - even in the bi-factorial framework - provided that all three sets in (4) can be characterised by scalar parameters, which explains direct applicability of the Rasch model for the assessment of CR systems

To illustrate the technique with real assessment data (though limited of course and not technical) we present here the results of the assessment of a group of 46 students of Technical University of Berlin that in 2006-2008 took the course "Advanced Internet Services" (lecture course 0432 L 746 OKS). As the written examination of the course all students were given 16 questions with four alternative answers per question. From the Rasch model viewpoint we interpret this experiment as series of 16 tests performed by 46 subjects. The ability parameter of a subject was computed as the percentage of correctly answered questions; the easiness parameter of each test was computed as the percentage of humans that provided correct answers. Despite the fact that each question had four alternatives our interpretation is dichotomous since we count only correct - incorrect outcomes. To simplify the outcome of the analysis we have analysed groups of subjects instead of single individuals; the subjects were grouped based on their equal abilities shown in the tests. Figure 23.a shows the distribution of relative easiness of all tests, Figure 23.b shows the distribution of relative abilities of groups of subjects; both scales are normalised.

Indeed in full conformance with the Rasch model the location of a particular subject on the common horizontal scale corresponds to the ability of that subject to answer correctly with 50% probability the question of the easiness that corresponds to the same common location on the scale; the easier the question the higher is the probability of correct answer for the given subject, and vice versa. Interestingly, the effect of grouping shows that a group has a higher potential for correct answers than the same set of subjects taken individually.

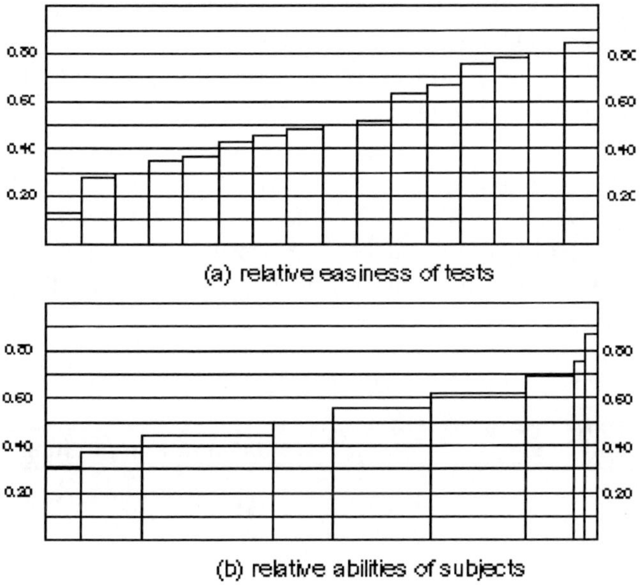

Figure 23. Normalised scales of factors.

Objective Monitoring Metrics

This section demonstrates that the distributed controller system is capable of reflecting cognition in both control loops in the implementation agnostic manner; meaning that regardless of cognition algorithm the distributed controller system can be used as a model of cognitive radio for the assessment of cognition algorithms.

Objective monitoring in our distributed controller system can be modelled as the sensing of a signal in generally noisy environment. The quality of objective monitoring depends on the amount of spent resources, such as time and energy. Spectrum sensing by CRS is the first task that is empowered by cognition; from the assessment viewpoint we are interested in metrics that will allow an object to enforce the policy triggered by the maximal likelihood (confidence) of sensed value to the target event (E_A). The purpose of cognition in the objective monitoring is to compute the confidence of the sensed event.

Let us consider the confidence (C) as a function of spent resource (R). We consider a confidence functions $\{C_i(R)\}$ that monotonously increase with the amount of resources spent for both sensing and cognition. Confidence in detecting the target event might grow from zero to the value denoted as Low (L), that is $C_i(R, L) = L$, and eventually proceed to the value denoted as High (H); that is $C_i(R, H) = H$. With the confidence above Low our objective monitor enforces cognition that helps to unambiguously determine the target event by various means - radio-specific as well as radio-agnostic. The confidence level Low can be characterised by the minimal amount of sensing resource that is needed to start learning. When our confidence function is between Low and High the cognition cycle is applied perhaps as a supervised sensing. We define in Table 3. the following primary cognition metrics for objective monitoring - likely typical areas and points of the confidence function.

Table 3. Primary cognition metrics (objective monitoring).

Acronym	Full Name	Meaning and Computation
MiSR	Minimal Sensing Resource	required to trigger the cognition; the smallest of all experienced C(R)
MxSR	Maximal Sensing Resource	allowed to be spent in the detection of the target event; configured upper bound
CSR	Cognition Start Range	the resource range between the minimal MiSR (the earliest cognition trigger) and the maximal MiSR (the latest cognition trigger); the difference between maximal and minimal values of MiSR for all C(R)
CFR	Cognition Finish Range	the resource range between the minimal MxSR (the earliest generation of the target event) and the maximal MxSR (the latest generation); the difference between maximal and minimal values of MxSR for all C(R)
MiCR	Minimal Cognition Range	the minimal resource range between MiSR and MxSR; difference for one function minimal for all C(R)
MxCR	Maximal Cognition range	the maximal resource range between MiSR and MxSR; difference for one function maximal for all C(R)

We use these primary metrics to engineer some derivatives useful for the establishment of the needed for the assessment relation between cognition in objective monitoring and cognition in subjective monitoring. The first derivative shown on Figure 24.a is the cognition envelope, it shows the boundaries of confidence maximally obtained for particular resources spent on cognitive sensing. The shape of this envelope within [MiSR,MxSR] is almost arbitrary since the envelope is only an instrument for further deliberations. The next derivative is shown on Figure 24.b, it is also instrumental and is termed cognition entropy. The cognition entropy is modelled after the binary entropy function known as Bernoulli trial since the goal of the objective monitoring is to detect with the high confidence the presence or absence of a single target event. In case more events need to be detected we easily cascade required sensors. We deliberately apply two entropy functions separately for Low and High confidence outcomes since - as it appears almost in all studies on cognition (see e.g. the survey [[40]] and [[41]]) - the process of cognition is non-linear in many respects.

The diagram Figure 24.b corresponds to the MxCR case of cognitive sensing, in which the detection of Low confidence requires on average more resources than the detection of the High confidence. For a typical CRS example this can be a case of the wider spectrum sensing at the beginning of the cognition cycle, in which a hint of an opportunity is detected that is followed by more focused sensing around the found hint. Contrary to the MxCR case the diagram Figure 24.c shows the MiCR case - initial opportunity hint is on average sensed faster than the more detailed sensing that follows. Since sensing in CRS is a continuous activity, or at least spreads several consecutive sensing intervals, it is important to consider the dynamics of cognitive sensing.

The goal of cognitive sensing is to learn from experience, the experience in our case is given by sensing parameters and by the achieved results in a number of consecutive intervals. The confidence function $C_4(R)$ might show that the cognition process losses the obtained confidence level; for example an already detected opportunity hint might be broken by suddenly sensed collision in place of the hint. When the Low confidence is lost after the CSR ends and not obtained again before the CFR starts it is perhaps reasonable for objective monitoring to consider the target event as not detected in this cognition cycle and to start the next cognition cycle with the Low confidence value as the outcome of the previous cycle. This reasoning - discrete cognition cycles but continuous learning - is shown in the diagram Figure 24.d; one can consider that continuous learning is achieved by a sliding window of the (in this case) size MxCR that relates the outcomes of the consecutive cognition cycles.

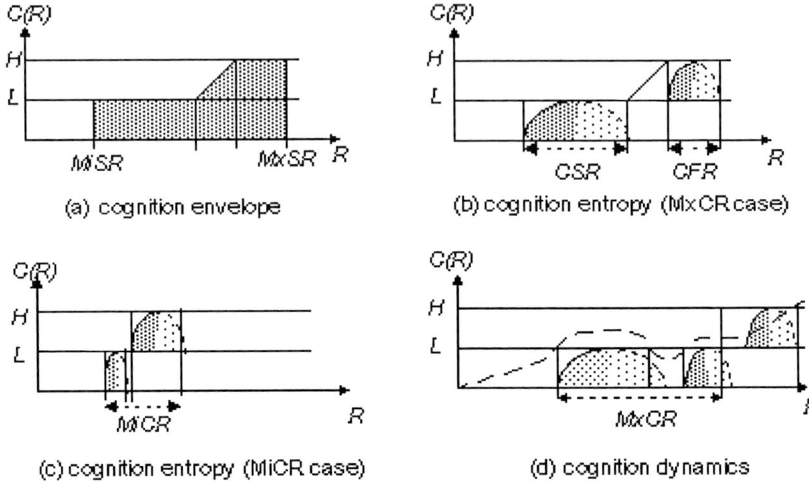

Figure 24. Secondary cognition metrics.

The goal of assessment of cognitive sensing is to evaluate how well the competence gained in the process of learning is serving the control purpose. The control purpose in our distributed controller system is given differently to different controllers; better to say that control purpose is represented as different state data within different controllers. To the objective monitoring the control purpose state is given as a set of conditions (events) enforcing the authorisation policies set on the object part of CR; to the object the control purpose state is given as a combination of authorisation policies and behaviour requests from the subject. When all three representations match in the policy domain (subject requests only those behaviours, which are authorised by needed events detected in the sensed environment) and in the time domain (events are still valid, both authorisation and obligation policies are still enforced) the control purpose is obviously met. However CR systems must operate in uncertainty; following the above reasoning we could define the two types of uncertainty- policy uncertainty that in our system appears as the mismatch (conflict) between policies, and time uncertainty appearing as the lack of resources for complete cognition.

Instead, since our distributed controller system acts as a composition of subject and objects we apply the methodology inspired by the Rasch model in the attempt to eliminate perhaps multiple and peculiar differences of objective monitoring in different spectra. Each objective monitor has its individuality; all the individualities must be abstracted for a subject to make best match to the control purpose. Following the cognition dynamics introduced in Figure 24.d we shall consider a sequence of adjacent cognition cycles as a sequence of tests and apply to the outcomes of these tests the specific objectivity framework, which is briefly introduce above.

Assessment of Process Correctness

The assessment of process correctness is the assessment of how well the control purpose is met by the composed behaviour of our distributed controller system that represents a single policy domain, in which all subjects and objects are configured with similar policies. From this viewpoint we argue that the purpose of cognition can be formulated as continuous consensus building between pairs of controllers. We attempt now to demonstrate that regardless of the implemented cognition algorithm our distributed controller system is the correct model for the assessment of those algorithms. Following the cognition dynamics introduced in Figure 24.d we shall consider a sequence of adjacent cognition cycles as a sequence of tests and apply the specific objectivity framework to their outcomes.

Cognition dynamics in our settings is constrained by the amount of resources that can be used for the objective sensing and by the allowed sequences of major cognition phases. Major cognition phases are as follows: at boostrapping the level of confidence in detecting the target event is Uncertain (*U*); cognition then can proceed to the Low (*L*) level, and from Low to High (*H*), or eventually from High to Low, and from Low to Uncertain as shown by the Petri net at Figure 25.a. The seven transitions shown on the Petri net are the only allowed in our model; this is the constraint on possible sequences of cognition phases; the transitions between phases can be clustered as learning transitions (*LUL*, *LLH*), forgetting transitions (*FLU*, *FHL*), and keeping the confidence level transitions (*KU*, *KL*, *KH*).

For each place in the Petri net the following holds: if neither learning nor forgetting transition happens within the resource constraints then the keeping transition fires, which marks the start of a new cognition cycle with the previous confidence level, otherwise (if either learning or forgetting happens) the new cognition cycle starts with the newly obtained level of confidence. These dynamics can be observed with the initial fragment of the reachability tree of the Petri net shown on Figure 25.b: horizontal lines labelled *U*, *L*, and *H* represent confidence levels; the resource consumption is assumed increasing (not necessarily linearly; the linear form is used for simplicity) from left to right; we draw a thick line starting at one confidence level and reaching the neighbouring confidence level to depict the transition between the two levels. The reachability tree fragment is shown for the initial marking $\{U, L, H\} = \{1, 0, 0\}$, the tree continues infinitely as a regular pattern with two arbitrary transitions from phases U and H, and with three arbitrary transitions from the phase L.

We use Figure 25.b to depict the constraints on the four transitions that all but the first one are shown as happening in some arbitrary order, which is yet constrained by the inequalities (7)

$$R_{UL} \le \frac{CSR}{2} + MiSR \quad R_{LH} - R_{UL} \le \frac{CFR}{2} \quad R_{LU} - R_{UL} \le \frac{CSR}{2} \quad R_{HL} - R_{LH} \le \frac{CFR}{2} \qquad (7)$$

These inequalities simply mean that objective monitoring has its unique internal structure fully defined by cognition phases and parameters. This structure can itself be exploited within the cognition to improve its results, however it is mostly important in the assessment of CR systems. We choose deliberately the resource constraints for learning and forgetting transitions to be 50% of respective average resource ranges since this corresponds to the maximum of cognition entropy for the MxSR case (Figure 24.b), meaning that the worst case ability parameter of an objective monitor (z_n in the expression (3)) is set to $z_n = 0.5$.

CR systems are expected to adapt to the changes in the environment, which requires the adaptation of the cognition itself. In application to the objective monitoring this means that the reachability tree of the Petri net model - and each individual branch of it - will be more and more departing from the regular pattern shown in Figure 25.c as a trellis of thin lines; the trellis is a product of arbitrary transition between confidence levels in a slotted (equal resource R for each cognition cycle) sensing. We expect that good CR sensing with the increase of knowledge will tend to keep the High confidence level for longer periods, it will be returning to Uncertain level only in a few really critical cases, while transiting the Low level relatively quickly in both directions. The expected (good) cognition process is shown as the thick line in Figure 25.c; the process is shown to keep the High confidence level almost 70% of the first seven after bootstrapping cognition cycles, which can be defined as the weight of the High confidence phase in the dynamic cognition process following the usual computation of the ratio of cognition slots with High level to all elapsed slots, subject to appropriate ageing of "old" slots.

Obviously, out of the three cognition phases only two - High and Low confidence in detecting a target event are reasonable to weight as discussed above proportionally to the obtained and yet valid knowledge. We believe that the two types of knowledge, respectively for Low and for High confidence need to be distinguished since they apply significantly different CR sensing strategies; for example, following [[41]] the strategy in the Low confidence phase might require larger entropy.

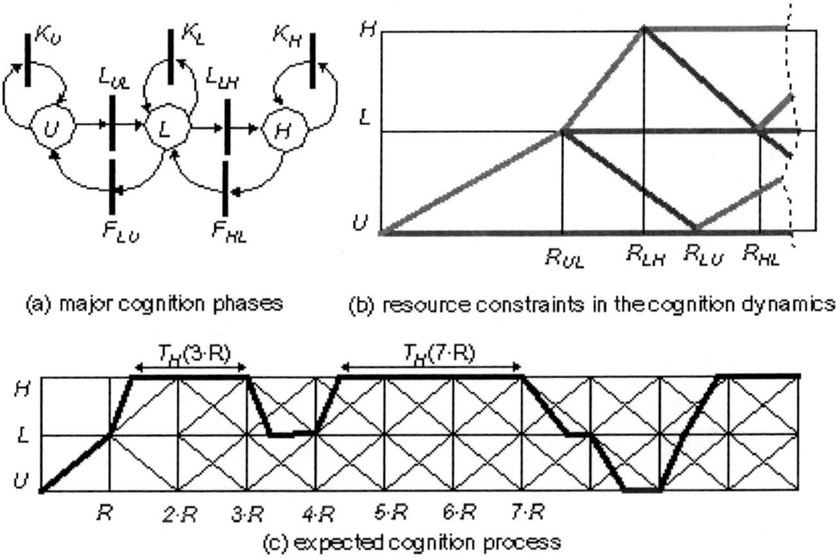

Figure 25. Cognition process (objective monitoring).

It is reasonable to share the weights of cognition phases of objective monitor with the subjective monitor. Since we cannot expect that CRS will operate always in High confidence of all target events our distributed controller system makes decisions in uncertainty; more precisely, the uncertainty characterises the behaviour of both monitors, while subject and object are making very certain decisions. Subject's decision is to select for the enforcement that obligation policy that minimises the risk; object's decision is to apply the right authorisation policy to the requested action. Both subject and object might have not enough certain information about events and conditions that prefix their respective policies; the decision in this case is to select the right - the one minimising risk - policy and enforce it as if the policy if the fully qualified one. Again, in both cases missing (uncertain) part of required information (i.e. events and conditions) is substituted by the prediction. An object uses cognition and own entropy to predict the confidence level of target event; a subject uses prediction to quantify the risk of the decision taken.

It is clear now that the efficiency requires that both types of predictions need to be made in consensus; even more than that - it appears that the consensus mechanism is helpful for the both types of predictions. As a background motivation for the consensus building as the right prediction technique let us recall Delphi method proved to be very successful exactly in the area of prediction; let us also recall the phenomenon of wise crowds, which points in the same direction. Even in our small experiment with the students we have observed that out of 64 possible choices in the test the students have mostly converged on 28 alternatives, which include 15 out of 16 correct choices.

Indeed, our distributed controller system is well suited for building a consensus between pairs of controllers since they represent a single policy domain. Objective and subjective monitoring though performed in different realms is the coordinated prediction of the best behaviour in a given situation. Objective monitors perceive the spectrum situation as confidence level of detecting of target events; subjective monitor perceives the decision situation as a collection of said confidence levels. Cross-adjustments of parameters between pairs of controllers is the consensus that brings the CRS to the converged behaviour choices.

Conclusion

Cognitive Radio Systems and the networks that they facilitate are fundamentally different from the wired Internet. First, they exhibit an order of magnitude larger adaptation space, in which due to the nature of radio communication they will need to adapt themselves and by this shall facilitate the adaptation of the entire network and its features, multiplexing being one of those.

Similar to the transition of transmission from circuit switching to packet switching in wired and wireless, we observe the management transition from stovepiping to policy-based management. This transition is fairly in the middle: we already know what machinery is needed for the PBM. This machinery will make significant part of network housekeeping and is present currently in information models, ontologies, specifications, policy and configuration languages, etc. - in other words this future is present today in the form of theories.

What is the next step? If one wants to keep it short: the network needs to learn these theories!

The longer answer is provided by our examples of self-learning policy, of learning own entropy, etc. In all these example all the enabling policy machinery disappears.

Imagine, the network shall learn these theories, but what if some of these theories are wrong or incomplete? Fundamentally important is to be able to assess cognitive networks, be able to verify their process correctness. During such understanding we also verify our policy theories.

Acknowledgements

This work is performed in project E3 which has received research funding from the Community's Seventh Framework programme. This chapter reflects only the authors' views and the Community is not liable for any use that may be made of the information contained therein. The contributions of colleagues from E3 consortium are hereby acknowledged.

References

[1] Dr. Joseph Mitola III, Challenges for integrating business logic with domain-specific technology-specific policy languages, IEEE P1900,5 Committee working document, 2008, on-line at http//grouper.ieee.org

[2] Lars Berlemann & Stefan Mangold. Cognitive Radio and Dynamic Spectrum Access, on-line at http://eu.wiley.com/WileyCDA/

[3] Wikipedia contributors, "Cognitive radio," Wikipedia, The Free Encyclopedia, http://en.wikipedia.org/w/index.php?title=Cognitive_radio&oldid=273616945 (accessed March 12, 2009)

[4] James O. Neel, Jeffrey H. Reed & Allen B. MacKenzie. "Cognitive Radio Performance Analysis", pp. 501-579, Chapter 15 in "*Cognitive Radio Technology*", Bruce Fette (Ed.), Newnes/Elsevier, 2006, 622 p.

[5] Graca Carvalho. Oral communication at the FET concertation meeting in response to the question "What is the single much desired add-on to the Internet technology?", Brussels, 2004

[6] S. Floyd. The RED page, on-line http://www.icir.org/floyd/red.html

[7] Wikipedia contributors, "Random early detection," Wikipedia, The Free Encyclopedia, http://en.wikipedia.org/w/index.php?title=Random_early_ *detection&oldid=151998484 (accessed October 4, 2007)*

[8] K. Kilkki. Differentiate Services for the Internet, *Sams, June 1999*, 384 p.

[9] Bruce Fette. SDR Technology Implementation for the Cognitive Radio, in Proceedings of the FCC Cognitive Radio Workshop, 2003

[10] Paul Baran, ODC, Memorandum RM-3420-PR, August 1964, Rand Corporation

[11] Paul Baran. ODC, 1964, v.1. RM-3420-PR, Chapter 4

[12] John Strassner. "How Policy Empowers Business-Driven Device Management", *IEEE, Proceedings of POLICY '02*, doi: 10.1109/ POLICY.2002.1011311

[13] C. de Laat, G. Gross, L. Gommans, J. Vollbrecht & D. Spence. "Generic AAA Architecture", RFC 2903, 2003, IETF, on-line: http://www.ietf.org/rfc/ rfc2903.txt?number=2903

[14] Seventh IEEE International workshop on Object-Oriented Real-Time Dependable Systems, San Diego, California, 7-9 January, 2002

[15] David C. Plummer. An Ethernet Address Resolution Protocol, IETF, RFC 826, November 1982

[16] Wikipedia contributors, "NGOSS," Wikipedia, The Free Encyclopedia, http://en.wikipedia.org/w/index.php?title=NGOSS&oldid=285547880 (accessed April 22, 2009).

[17] B. Aiken, J. Strassner, B. Carpenter, I. Foster, C. Lynch, J. Mambretti, R. Moore & B. Teitelbaum. Network Policy and Services: A Report of a Workshop on Middleware, IETF, Request for Comments: 2768, February 2000

[18] Wikipedia contributors, "Information model," Wikipedia, The Free Encyclopedia, http://en.wikipedia.org/w/index.php?title=Information_ model&oldid=258654752 (accessed January 23, 2009).

[19] Virgil Gligor. University of Maryland, National Security Award 2006, Invited talk at The 3rd Annual VoIP Security Workshop, Berlin, Fraunhofer FOKUS, 01.JUN.2006

[20] John Strassner "How Policy Empowers Business-Driven Device Management", *IEEE, Proceedings of POLICY '02*

[21] Paul Baran. Visions of the 21st Century Communications: Is the Shortage of Radio Spectrum for Broadband Networks of the Future a Self Made Problem?, Keynote Talk Transcript, 8th Annual Conference on Next Generation Networks Washington, DC, November 9, 1994, on-line http://www.winlab.rutgers.edu/~crose/baran.html

[22] Wikipedia contributors, "Imprinting (psychology)," Wikipedia, The Free Encyclopedia, http://en.wikipedia.org/w/index.php?title=Imprinting_ (psychology)&oldid=294325043 (accessed June 4, 2009).

[23] John Strassner. "Introduction to DENng for PanLab II", ACF, 2008-2009, tutorial given 21.01.2009 in Fraunhofer FOKUS

[24] M. Fisher & P.Mckee. Policy Based Management Of Large Distributed Systems, BTexact paper, CaberNET

[25] Farhad Arbab. "Abstract behavior types: a foundation model for components and their composition", in Formal methods for components and objects pragmatic aspects and applications, Volume 55 , Issue 1-3 (March 2005) Pages: 3 - 52 ISSN:0167-6423

[26] Jan H. van Schuppen. "Problem 4.4 Decentralised control with communication between controllers", in V.D. Blondel, A. Megretski (Eds), "Unsolved Problems in Mathematical systems and Control Theory", Princeton University Press, 2004

[27] Michael Smirnov. QoS ad hoc Internetworking: Dynamic Adaptation of Differentiated Services Boundaries, in Next Generation Networks, LNCS 1938, pp. 186-195

[28] Ivana Dusparic & Vinny Cahill. Distributed W-Learning: Multi-Policy Optimization in Self-Organizing Systems, in the Proceedings of Third IEEE International Conference on Self-Adaptive and Self-Organizing Systems, San Francisco, California, September 14-18, 2009

[29] S. Floyd. The TCP-Friendly Website, on-line http://www.icir.org/floyd/*tcp_friendly.html*

[30] N. G. Duffield. Exponential bounds for queues with Markovian arrivals Journal Queueing Systems ISSN 0257-0130 (Print) 1572-9443 (Online) Issue Volume 17, Numbers 3-4 / September, 1994

[31] Michael E.Young & Edward A. Wasserman, Stimulus Control in Complex Arrays, chapter in the book Robert G. Cook (Ed.) Avian Visual Cognition, Sept. 2001, Department of PsychologyTufts University In cooperation with Comparative Cognition Press, on-line http://www.pigeon.psy.tufts.edu/avc/toc.htm

[32] The E2R II Flexible Spectrum Management (FSM) Framework and Cognitive Pilot Channel (CPC) Concept Technical and Business Analysis and Recommendations", E2R II White Paper, November 2007, 52 p., on-line at http://e2r2.motlabs.com/whitepapers

[33] XG Working Group. "XG Policy Language Framework", Request for Comments, v.1.0, April 16, 2004, prepared by BBN Technologies, Cambridge, Mass, USA, on-line at http://www.ir.bbn.com/projects/xmac/index.html

[34] B. Lane. "Cognitive radio", in Cognitive Radio Technologies Proceeding (CRTP) ET Docket No. 03-108, on-line at http://www.fcc.gov/oet/cognitiveradio/

[35] XG Working Group, "The XG Architectural Framework", Request for Comments, v.1.0, prepared by BBN Technologies, Cambridge, Mass, USA, on-line at http://www.ir.bbn.com/projects/xmac/index.html

[36] Mikaël Beauvois. "Brenda: Towards a Composition Framework for Non-orthogonal Non-functional Properties", Book chapter in Distributed Applications and Interoperable Systems, LNCS Vol. 2893/2003, pp. 29-40, on-line at http://www.springerlink.com/content/qda3mkrxhlaxmpee/

[37] Ram Ramanathan. "Wireless Research: Challenges and Future Directions", Industry Panel: NSF NETS Meeting, December 2007, on-line at http://www.ece.iit.edu/~nsfwmpi/abstracts/Event_Speakers/Ramanathan_nets-panel.ppt

[38] Simon Dobson. "Facilitating a well-founded approach to autonomic systems", in the proceedings of the 5th IEEE Workshop on Engineering of Autonomic and Autonomous Systems (EASe 2008)

[39] Georg Rasch. "On Specific Objectivity: An Attempt at Formalizing the Request for Generality and Validity of Scientific Statements", 1977, on-line at http://www.rasch.org/memo18.htm

[40] D. Vernon, G. Metta, & G. Sandini. "A Survey of Artificial Cognitive Systems: Implications for the Autonomous Development of Mental Capabilities in Computational Agents", IEEE Transactions on Evolutionary Computation, special issue on Autonomous Mental Development, Vol. 11, No. 2, pp. 151-180, 2007. on-line at http://www.vernon.eu/publications.htm

[41] R. V. Belavkin & F. E. Ritter. "The use of entropy for analysis and control of cognitive models". In F. Detje, D. Doerner, and H. Schaub, editors, Proceedings of the Fifth International Conference on Cognitive Modeling, pages 21--26, Bamberg, Germany, 2003. http://citeseer.ist.psu.edu/belavkin03use.html

In: Cognitive Radio ISBN 978-1-60876-604-8
Editors: P.D. McGuire et al, pp. 203-236

Chapter 7

TOWARDS COGNITIVE RADIO FOR EMERGENCY NETWORKS

Qiwei Zhang, Fokke W. Hoeksema, Andre B.J. Kokkeler and Gerard J.M. Smit*
Department of Electrical Engineering, Mathematics and Computer Science, University of Twente, P.O. Box 217, 7500 AE Enschede, The Netherlands

Abstract

Large parts of the assigned spectrum is underutilized while the increasing number of wireless multimedia applications leads to spectrum scarcity. Cognitive Radio is an option to utilize non-used parts of the spectrum that actually are assigned to primary services. The benefits of Cognitive Radio are clear when used in emergency situations. Current emergency services rely much on the public networks. This is not reliable in emergency situations, where the public networks can get overloaded. The major limitation of emergency networks is spectrum scarcity, since multimedia data in the emergency network needs a lot of radio resources. The idea of applying Cognitive Radio to the emergency network is to alleviate this spectrum shortage problem by dynamically accessing free spectrum resources. Cognitive Radio is able to work in different frequency bands and various wireless channels and supports multimedia services such as voice, data and video. A reconfigurable radio architecture is proposed to enable

*E-mail address: G.J.M.Smit@utwente.nl

the evolution from the traditional software defined radio to Cognitive Radio.

Keywords: Cognitive Radio, Emergency Networks, Spectrum Sensing, OFDM, Reconfigurable Platform

1. Introduction and Background

Recent studies show that most of the assigned spectrum is underutilized. On the other hand, the increasing number of wireless multimedia applications leads to a spectrum scarcity. Cognitive Radio ([1], [2]) is proposed as a promising technology to solve the imbalance between spectrum scarcity and spectrum underutilization. In Cognitive Radio, spectrum sensing is done in order to locate the unused spectrum segments in a targeted spectrum pool and use these segments optimally without harmful interference to the licensed user. A lot of research projects have started to prove the concept of Cognitive Radio and to demonstrate its feasibility. In the Adaptive Ad-hoc Freeband (AAF) [3] project, we design a Cognitive Radio based wireless ad-hoc network for emergency situations. Although the AAF project addresses Cognitive Radio in a complete fashion from physical layer to networking issues, this book chapter mainly focuses on the physical layer related issues including: spectrum sensing, baseband transmission and physical layer reconfigurability.

In section one, we give background information on the AAF project followed by an introduction to Cognitive Radio and related research projects. Section two is dedicated to spectrum sensing which is an essential functionality of Cognitive Radio. An OFDM based Cognitive Radio system is discussed in section three. In section four, we propose a reconfigurable platform to support Cognitive Radio.

1.1. Requirements for Emergency Networks

Current day emergency services rely for data communications on public radio networks like GPRS. Sometimes in disaster situations, even GSM is used for voice communication between relief workers. However, in case of emergency the public networks may get overloaded. So, the use of generally available public networks is not considered to be reliable enough for emergency situations. Moreover, GSM/GPRS is an infrastructure based network and therefore highly susceptible to disasters in small and medium sized urban areas. A real-life example is the S.E. Fireworks disaster in Enschede, the Netherlands (May 2000), where a fireworks depot exploded and destroyed a large part of the city, killing 23 people and injuring more than a thousand. In the first hectic hours after the last of the large explosions, the municipal disaster command in particular had great difficulty to gain an overview of the extent of the disaster, and of the situation in the disaster area. Fire brigade, police and relief workers in the medical chain experienced severe communication problems, both internally and with each ther, because transmission equipment appeared not to be working, or was functioning inadequately.

In such a scenario it is vital that communication between relief groups is established as quickly and as easily as possible. The fact that the infrastructure-based networks in such an area may be destroyed raises the need to communicate using infrastructure-less wireless technologies. So, a Radio Access Network (RAN) or some type of cellular network (GSM, UMTS) will not do. Access to the core telecommunications network will only be available at the periphery of the disaster area where the damage to the infrastructure is less severe. The area of operation of the emergency network is confined to the disaster area, which is, for example, no larger than a circle with a 2.5 kilometer radius (comparable to the Enschede case). An ad-hoc networking approach will allow the relief groups to enter the disaster area and communicate with each other quickly. With slightly more effort, the wireless ad-hoc network can be hooked up to the original core telecommunications network almost instantly. The relief network must be able to handle multimedia traffic service to its users. A reliable real-time voice communication is especially important in emergency situations. Data messages like location coordinates, sensor data and even the health status of the rescue workers are also needed. Pictures and video can be used for surveillance and locating objects. To support multimedia services, different constraints on QoS have to be met. Real-time voice and video are sensitive to delay and jitter while data messages have high requirements on data loss and error rates. The voice and video provide more or less predictable network load due to their streaming behavior but data traffic is rather bursty in nature. In short, the network must be able to handle a wide variety of multimedia signals and has to deal with large, possibly unpredictable amounts of data. The large amounts of data require a large amount of radio resources.

There are also public safety networks like C2000 which is based on TETRA (see table 1).

Table 1. Radio parameters of the TETRA System.

Parameter	Value
Frequency Range	380 – 400 MHz
Channel Spacing	25 kHz
TX Power (mobile)	1 – 3 W
Channel Access method	TDMA
Modulation method	$\pi/4$-DQPSK
Channel bit rate	36 kbit/s
Maximum data rate (gross bit rate)	28.8 kbit/s
Net data rate, non-protected:	n x 7.2 kbit/s
Low-protected	n x 4.8 kbit/s
High-protected	n x 2.4 kbit/s
	(n = 1, 2, 3 or 4)
Speech coding	A-CELP, 4.567 kbit/s
Range Rural	± 14 km
Range Suburban	± 4.5 km

One may observe that there are limited data communication facilities available in C2000. Also this network is infrastructure based (there are fixed base stations) and thus susceptible to the type of disasters we conceive.

The state-of-the-art systems lack capabilities (e.g. in offered data rates or multimedia traffic support) and are not disaster proof. The other major drawback of current emergency networks is their spectrum scarcity [4] because current emergency networks are assigned with a limited spectrum and fixed bandwith. As we mentioned earlier, the large amounts of multimedia data in the emergency networks require a large amount of radio resources. One band can easily get congested due to heavy traffic, which makes it inadequate for emergency use. If several fragmented bands are assigned to emergency use, the interoperability and the lack of standards will become another problem [5]. Therefore to alleviate this spectrum shortage problem, a radio which dynamically accesses free spectrum resources turns out to be an interesting solution. This approach is called Cognitive Radio according to Mitola [1], who first came up with the concept. He defined Cognitive Radio as a radio that can change its transmission based on interaction with its environment. This definition can be very broad, including knowledge of services, user behaviour and spectral usage. The AAF project focuses on two key elements, i.e. searching for under-utilised spectrum, and (rapidly) adapting transmission settings accordingly.

The first step in the project is to focus on identification of free resources in the frequency domain by spectrum sensing. Subsequentially we identify an OFDM-based system which is theoretically optimal in approaching the Shannon capacity in the segmented spectrum by sending at different rates and powers over each subcarrier. We believe that the capacity to nullify individual carriers poses interesting opportunities for cognition, as was also observed in [6]. Cognitive Radio has to operate in different frequency bands, combat various negative effects of wireless channels and support various multimedia services. Therefore Cognitive Radio needs physical layer reconfigurability. So, we propose a heterogeneous reconfigurable platform to enable the evolution from the traditional software defined radio to Cognitive Radio.

1.2. Cognitive Radio

The idea of Cognitive Radio was first presented by Joseph Mitola III in his paper [7], where he proposed that Cognitive Radio can enhance the personal wireless service by a radio knowledge representation language (RKRL). This language represents knowledge of radio at all aspects from transmission to application scenarios in such a way that automated reasoning about the needs of the user is supported. Explained in a simple way, Cognitive Radio is able to autonomously observe and learn the radio environment, generate plans and even correct mistakes. A comprehensive conceptual architecture of Cognitive Radio was later presented in [1], where Cognitive Radio was thought as a final point of the software-defined radio platform evolution: a fully reconfigurable radio that changes its communication functions depending on network and/or user demands. Mitola's work covered interesting research subjects in multiple disciplines such as wireless communication, computer science and cognitive science.

On the other hand, the radio regulatory bodies in various countries found that most of the radio frequency spectrum was inefficiently utilized. In November 2002, the Federal Communications Commission (FCC) in the United States released a report [8] aimed at improving the management of spectrum resources in the US. The report concluded that the current spectrum scarcity problem is largely due to the strict regulation on spectrum access. The spectrum measurements conducted by the FCC indicated that only small portions of the spectrum are heavily used while other frequency bands are either partially used or unoccupied most of the time. So, spectrum utilization can be improved by making it possible for a secondary user (who doesn't have the license for spectrum) to access the spectrum which is not occupied by the licensed user (primary user). This secondary user has the awareness of the spectrum and adapts its transmission accordingly on a non-interference basis. This spectrum access and awareness scheme is referred to as Cognitive Radio by the FCC, which is a narrower definition compared with the original concept brought up by Mitola.

1.3. Projects Related to Cognitive Radio

Since the concept of Cognitive Radio first appeared, research activities have started around the world. Recently, Cognitive Radio became a very hot topic due to its impact on future spectrum policy which could fundamentally change the current status of radio communication. We mention a few research projects related to Cognitive Radio.

In Berkeley Wireless Research Center (BWRC), a dedicated Cognitive Radio Research (CRR) project is in progress. A white paper on Cognitive Radio has been produced [9] earlier and recently they reported a real time Cognitive Radio testbed for physical layer and link layer experiments [10]. Their motivation is to improve the spectrum utilization by opportunistic use of the spectrum, which is the same as the FCC's initiative. They treat the subject in a complete fashion: from physical layer issues to MAC layer issues and from analog frontend to computing platform supporting baseband processing. It is interesting to look into their study on spectrum sensing. From the system perspective, they study some basic considerations: the link budget of the sensing; the effect of noise on sensing; the cooperation of sensing nodes. From the signal processing perspective, a comparison study is made on different sensing techniques. Some hardware considerations related to sensing are also proposed see [11].

Spectrum pooling [6] is investigated by Jondral from the University of Karlsruhe. The basic idea is that a secondary user can dynamically access the licensed band by switching on and off OFDM subcarriers to avoid interference to the licensed user (primary user). However, the spectrum power leakage problem in the FFT based traditional OFDM systems could cause potential interference to the licensed system. They also observed this in [12], where two counter measures are proposed: spectrum shaping and switching off the carriers adjacent to the licensed user (primary user). There are other challenges for spectrum pooling like detection of spectrum access and synchronization, see [6].

The Cognitive Radio project [13] at Virginia Tech. does not specifically aim to improve spectrum utilization. This project is based on the observation that Cognitive Radio distinguishes itself by awareness and learning. In [14], a genetic algorithm based cognitive engine is proposed to learn its environment and respond with an optimal adaption. This approach to Cognitive Radio is more or less similar to the original concept of Mitola.

The European Union 6th framework End-to-End Reconfigurability (E2R) [15] project brings together major European players in the domain of reconfigurability, software defined radio and cognitive radio. The key objective of the E2R project is to devise, develop, trial the architectural design of reconfigurable devices and supporting system functions for users, application and service providers, operators, and regulators in the context of heterogeneous systems. Although the project does not specifically address Cognitive Radio, dynamic spectrum allocation and evolution from software defined radio to Cognitive Radio has been envisioned.

In parallel with the ongoing research projects around the world, international standardization organizations also have proposals to improve the spectrum utilization. An example is IEEE 802.22 [16], which is a new standard for a cognitive point-to-point (P2P) air interface for spectrum sharing with television bands. Television channels are very suitable for cognitive radio because they have a relatively unique spectrum signature that is easy for a cognitive radio to identify. The signals are also rigidly assigned to 6-MHz-wide channels with fixed center frequencies.

In the scope of our AAF project, we focus on the spectrum awareness and access. The objective is to demonstrate a Cognitive Radio system for emergency networks. Each radio node works in an ad-hoc based network, the node adopts the AAF protocol stack which is consistent with the five-layer protocol reference model (physical layer, data link layer, network layer, transport layer and application layer). The physical layer is the heart of our study where the free spectrum is discovered. Whenever the free spectrum is found, the AAF system will create an infrastructure using this spectrum.

2. Spectrum Sensing

In this section an architecture of a generic sensing system is described which consists of a spectral analysis system and a decision system. As an example, we present a sensing system that uses energy detection in the frequency range between 400−500 MHz. For disaster relief networking this range is interesting because of propagation conditions, power requirements and antenna dimensions.

A brief analysis into the gains to be expected by collaborative scanning, a form of distributed detection, completes this section.

2.1. Sensing System Architecture

In Figure 1 a functional architecture of a sensing system is presented. Basically the system consists of two parts: a spectral analysis system and a decision system. An option for the spectral analysis system is to use energy detection (power detection). In this case the scanning system works on any signal, however, without using all available knowledge of signal properties of the primary user. Another option is to use feature establishment - one identifies well known (deterministic) signal features of the primary user's signal like carrier waveforms or pilot tones. As, for instance in broadcast situations, the primary user wants to be heard, it is expected that especially in bad SNR conditions feature establishment may outperform energy detection [17]. In this paper we focus on energy detection, as it is applicable to *any* primary user signal and possibly less energy consuming than feature detection.

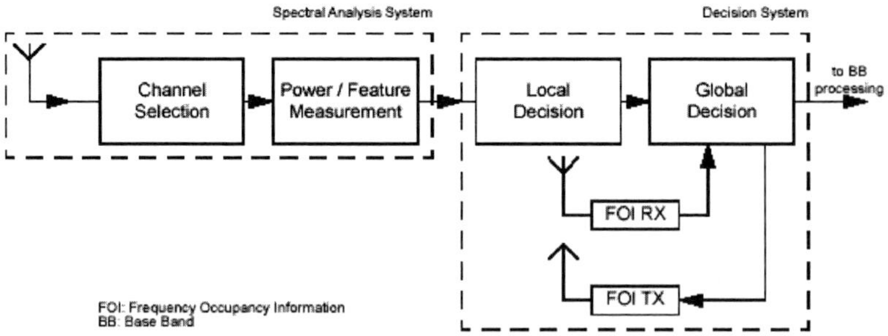

Figure 1. Functional Architecture of the sensing system.

A single scanning node needs to decide whether or not the band under consideration is empty or not, hence it takes a local decision. The quality of this decision (in terms of detection probability P_d and False Acceptance Rate (FAR)) may be hampered by two issues:

Multipath fading Signals that travel different signal paths may or may not add coherently at the receiver. As a result, small changes in location (relative to the wavelength λ) impact detection quality.

Shadowing An individual sensing node may be blocked by an obstacle and may therefore not be able to detect the primary user. Shadowing may occur over an area of $5 - 500$ m, depending on the size of the object.

To overcome these problems, a collaborative scanning system can be used. In such a system Frequency Occupancy Information (FOI) is gathered by all individual scanning nodes and disseminated to the nodes participating in the secondary-user network using a special signalling channel for this purpose. The properties of this FOI dissemination channel (like its bandwidth, SNR, data rate, MAC protocol and delay), the independence of measurements and the number of nodes involved in the scheme all determine the gains in detection quality to be expected over locally-made-only decisions.

Leaving aside feature detection, the channel selection and power measurement part together form a spectrum analyser. Below we focus on alternative architectures for such a system.

2.2. Architectures of the Spectral Analysis System

We distinguish three architectures for the spectrum analysis part of a power-based sensing system: a scan-based architecture, a real-time FFT based architecture and an FFT-based architecture.

2.2.1. Scan-based Architecture

In Figure 2a, a scan-based architecture is presented. After a broad low-pass filter, a mixer driven by a local oscillator (LO) downconverts the signal to some intermediate frequency (IF). Subsequently the signal is anti-alias filtered, mixer products are removed and analog-to-digital conversion (ADC) is performed. In the digital realm a sample-and-hold type of filter is used and subsequently the energy in a particular frequency bin is computed. Instead of measuring frequency bins in parallel, they are measured sequentially, by changing the frequency of the LO and sweeping over the required frequency span (FS).

One critical design issue (especially for handheld nodes) is the power consumption of the AD converter. In order to compare the performance of AD converters a Figure of Merit (FoM) can be used that takes into account the converter's power dissipation P [W], its resolution (specified in Effective Number of Bits (ENOB)) and its sampling rate f_s [Hz], [18], [19]. This FoM is given by

$$\text{FoM} = \frac{P}{2^{\text{ENOB}} \cdot f_s} \ [\text{pJ} / \text{conversion-step}]. \tag{1}$$

in which a FoM of approximately 2 pJ / conversion-step is valid for today's ADCs. As dynamic ranges of as large as 80 dB can be observed for the 400 MHz-500 MHz frequency range ([20], [21, pp.77-84]) approximately $80 / 6 = 14$ bits ENOB are needed (we apply the well-known 6 dB/bit rule for ADCs), which severely limits the possible bandwidth that can be converted (say, to around $f_s/2 = 1 - 10$ MHz, so power $P = 66 - 655$ mW).

However, the main disadvantage of the sweeping approach is the huge loss of time resolution. First, it takes time for the filters to settle, which is inversely related to the resolution bandwidth (RBW). So the sweep frequency has to 'dwell' a certain amount of time in every frequency band. Also, the completion of one sweep takes an amount of time proportional to FS/RBW, so the sweep time is equal to

$$ST = k \frac{FS}{RBW^2} \tag{2}$$

in which k is a factor that depends on the filter type with typical values between 2 and 3, [22].

In Figure 2b the time-frequency diagram of the scan based approach is shown. In this figure the primary parameters involved in the scan-based method are indicated. The scanning starts one time t_s on frequency f_L. The scanning ends on t_e on frequency f_H. The complete scan time is ST, and the frequency span is FS. The Inter-bin frequency (IBF) is the distance between two adjacent frequency bins. The Inter-bin time (IBT) is the time it takes to complete a power measurement in a single time-frequency bin. The frequency bins are described by two parameters: The resolution bandwidth (RBW) and the dwell time (Td).

The gray bins are the time-frequency points that are 'seen' by the spectrum analyser. The bins that are not on the gray diagonal are just missed. Depending on the usage pattern of the primary user, this may or may not be a problem. In case of stable broadcast bands it seems less of a problem than in data-communication bands. In the example the number of bins is #points=8.

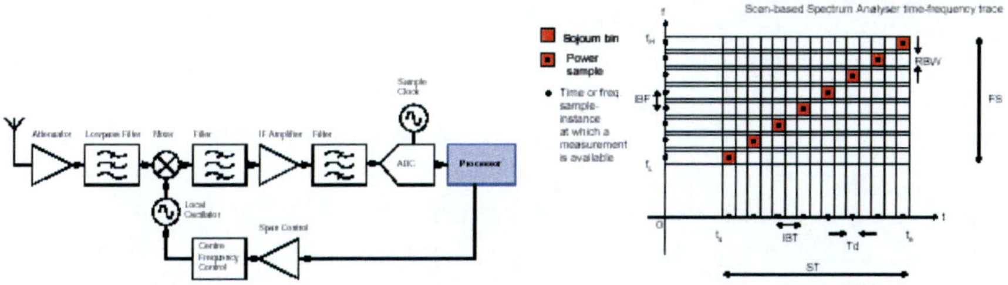

(a) Architecture. (b) Time-Frequency diagram.

Figure 2. Scan-based approach.

2.2.2. Real-time FFT Based Architecture

Another approach is to calculate the spectrum with a discrete Fourier transform (DFT) from time domain measurements instead of measuring it directly.

Since we want to perform the Fourier transform digitally we first have to convert the RF signal to baseband using a quadrature mixer, apply an anti aliasing filter, remove mixing products and apply analog-to-digital conversion. A digital signal $u[n]$ has a discrete Fourier transform $U[k]$ defined as:

$$U[k] = \sum_{n=0}^{N-1} u[n]\, e^{-j\,\frac{2\pi}{N}\,k\,n} \quad \text{with } k = 0, \ldots, N-1 \tag{3}$$

The conversion from time domain to frequency domain can be done with the very efficient FFT (Fast Fourier Transform) algorithm which takes only about $N \log(N)$ multiplications instead of the N^2 multiplications for a general DFT. In Figure 3a, the architecture of the FFT based method is shown. The front end is tuned on a fixed frequency. The channel select filter is replaced by a much wider filter that passes the whole frequency span. The selection of the frequency bins is now done by the FFT which processes all channels in parallel.

The primary parameters involved in the FFT-based approach are given in 3b. If we compare this figure with Figure 2b we see some important differences.

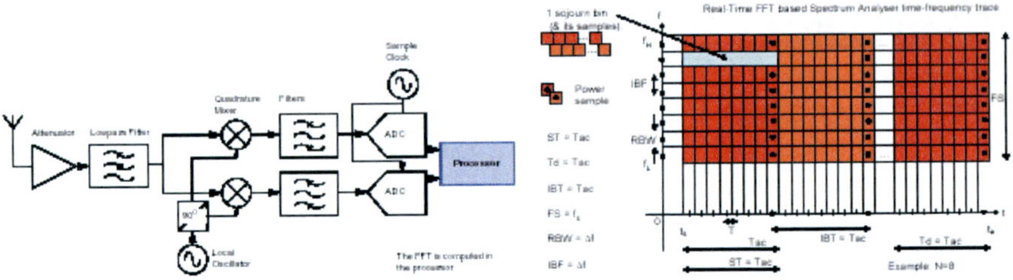

(a) Architecture.

(b) Time-Frequency diagram.

Figure 3. FFT-based approach.

First we see that now all bins are covered instead of just one diagonal. So there are no points in the time-frequency plane that are missed. We also see that the measurement results for all frequency bins become available at the same time, after the scan time ST, which is equal to the acquisition time if we neglect the processing time ($Tac = NT$, with T the sampling time). The frequency span is equal to the sample frequency $FS = f_s = 1/T$.

The real-time FFT based method has drawbacks too. The major disadvantage of the FFT method is that two ADCs is needed, which cause even more power-related problems as was the case for the scan-based architecture. For instance, an FS of 100 MHz cannot be accommodated. However, in the UHF/VHF region bandwidths of $1 - 10$ MHz seem feasible.

2.2.3. Scanning-FFT Based Architecture

A scan-based architecture is more energy-efficient compared to a real-time FFT based one, but considered too slow for our purposes. The real-time FFT based architecture is fast but too power intensive for a large frequency span, so the natural way to proceed is to combine both methods. One chooses a feasible FFT size (determined by the power requirements of the ADC) with a frequency span that is (thus) a fraction of the total span required. We then can use the FFT multiple times and adjust the centre frequency of the front end each time so that the whole span is covered. In Figure 4 the resulting architecture and the time-frequency diagram of this configuration are given (in the example, F=2 tuning steps of the LO are needed).

To provide some quantitative insight we provide an example. Assume we require a total frequency span of $FS = 100$ MHz with a resolution of $RBW = 10$ kHz, so 10000 bins. If we want to process this with a radix-2 FFT we need to

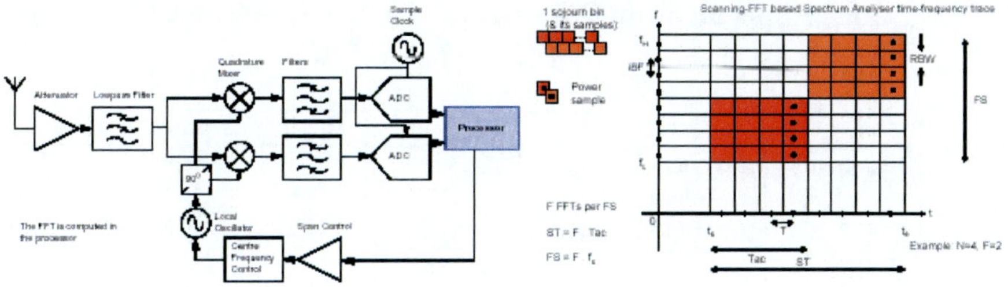

(a) Architecture.

(b) Time-Frequency diagram.

Figure 4. Scanning-FFT based approach.

sample and compute an 16384 points FFT every $Tac = 100\,\mu s$ (the first power of two larger than 10000). This makes up a substantial computational load.

If we use a scan-based architecture, according to equation (2) with k=2, it will take the sweep $2\frac{100\,.\,10^6}{(10\,.\,10^3)^2} = 2\,\text{s}$ to complete, which may be too long if we want to detect within a MAC frame duration (in the order of milliseconds).

If we use a scanning-FFT based architecture, we can partition the total span in 20 blocks of 5 MHz, which may be processed by calculating 512 points every $100\,\mu s$. And since the bandwidth is reduced by a factor 20, the sampling rate and thus the power consumption is also reduced by the same factor (see (1)).

Another benefit of a scanning-FFT based architecture is that the LO can interplay independently with a hopping baseband processing system, as was suggested in [23].

2.3. Specification Parameters for the Spectral Analysis System

The spectral analysis system can be described by the specification parameters of a spectrum analyzer as given in table 2. The meaning of the parameters is illustrated in figures 2b, 3b and 4b. Note that the parameters are not independent of one another and, moreover, the dependency depends on the chosen scan-system architecture (see [22], [24] and [25]).

As an example, consider a frequency range of interest from 400 MHz to 500 MHz. In order to see what current-day capabilities of energy-based sensing systems are, we investigate the capabilities of existing commercially available spectrum analysers (Agilent 4407B, Rohde & Schwarz FSP3 and Tektronics

Table 2. Specification Parameters for the Spectral Analysis System ([22], [24], [25]).

Aspect	Parameter	Unit	Variable	Remarks
Frequency	Frequency Span	[Hz]	FS	
	Resolution Bandwidth	[Hz]	RBW	
	Inter-bin Frequency	[Hz]	IBF	
Time	Sweep Time	[s]	ST	
	Dwell Time	[s]	Td	A.k.a. Sojourn time, S&H integration time, Time per bin
	Inter-bin Time	[s]	IBT	
Power	Dynamic Range	[dB]	ΔdB	
	Sensitivity	[dBm]	DANL	Displayed Average Noise Level
Memory	#points	-	Number of bins	

FSA3308A). Below, we present some parameter-*values* of these systems.

- The required frequency resolution of the scanning system can be taken from the Dutch National Frequency Plan [26] and from the Dutch frequency register [27]. In the range we studied (400 MHz - 500 MHz) a resolution bandwidth RBW=10 kHz seems sufficient, as the nominal channel spacing ranges from 12.5 kHz to 25 kHz (below 470 MHz) and from 200 kHz to 8 MHz above 470 MHz (TV bands).

- For a frequency span FS=100 MHz and a resolution bandwidth RBW=1 kHz the shortest sweep-time ST equals 8 s (using a Rohde & Schwarz FSP3 with #points=8000 and "FFT-filters" mode). This seems too long for MAC-frame durations (e.g. 2 ms for HiperLAN/2), but may be sufficient for course-grain measurements (e.g. with RBW's in the range of MHz's the sweep times (STs) become smaller).

- The studied spectral analysers do not provide high time-resolution in combination with high frequency resolution spectral measurement (the maximum real-time FFT bandwidth was 15 MHz (Tektronics FSA3308A), so FS=15 MHz).

- All studied spectrum analysers have a dynamic range of around 80 dB.

- The required time resolution of the scanning system depends on the utilisation pattern of the primary users.

As can be seen, a good tradeoff has to be found between time resolution, frequency resolution and frequency span. In case high frequency resolution is needed and the scanning needs to take place on the time scale of a MAC frame (milliseconds) a real-time FFT based architecture can be used, however with a limitation on the frequency span. A change in carrier frequency of the scanning system seems possible per MAC frame, so that the preferred architecture is a scanning-FFT based one.

2.4. Collaborative Scanning: Distributed Detection

In this secction we propose a decision system for the detection of a primary user signal. Suppose the decision system (see Figure 1) takes a local hard-decision (a yes/no one) on an energy-based measure. So, for the k-th frequency bin

$$E_k = |U[k]|^2 \tag{4}$$

is computed with $U[k]$ as in (3). Assume $U[k] = X[k] + W[k]$ in which $\{X[k]\}$ is the (complex-valued) primary user signal in the k-th frequency bin (which can be anything, with unknown distribution) and $\{W[k]\}$ is assumed to be (complex-valued) white Gaussian noise, independent of the primary user's signal. The expected value of the signal is $E[|U[k]|^2] = E[|X[k]|^2] + E[|W[k]|^2]$. Now, we formulate two hypotheses H_i:

H_0 : *no* primary user signal is present, so $|U[k]|^2 = |W[k]|^2$, and

H_1 : a primary user signal *is* present.

Subsequently we define a decision rule D:

$$D_k = 0 \text{ in case } E_k < \theta \text{ and } D_k = 1 \text{ in case } E_k > \theta$$

in which D_k is the decision made by the local decision system for the k-th frequency bin and θ is the detection threshold.

The quality of the detection can be expressed in two conditional probabilities, the detection probability P_d and the false alarm rate FAR. The probability of detecting a primary-user signal is given by $P_d \triangleq Pr[D_k = 1 \mid H_1]$, the probability of detecting noise instead of a primary-user signal is $FAR \triangleq Pr[D_k =$

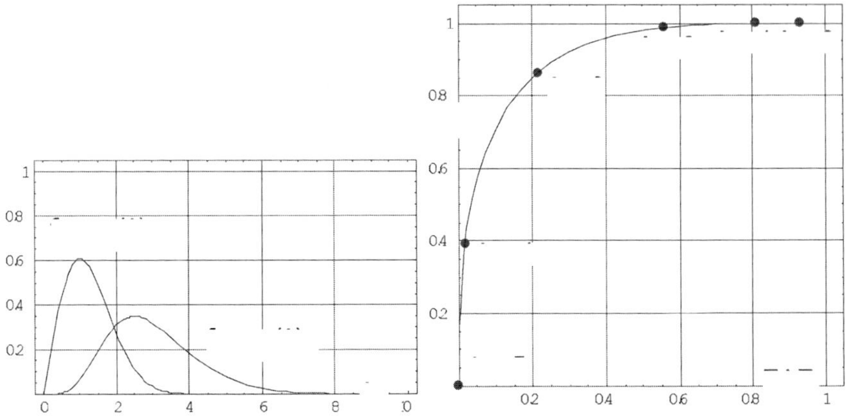

(a) Conditional pdfs $f_{E_k \mid H_0}$ and $f_{E_k \mid H_1}$. (b) ROC: P_d as a function of FAR.

Figure 5. Detection quality of a single node.

$1 \mid H_0]$. In case the probability density functions (pdfs) of E_k under both hypotheses are available, so $f_{E_k \mid H_1}(x)$ and $f_{E_k \mid H_0}(x)$ are known, the detection probability and false alarm rate can be computed as

$$\begin{aligned} P_d &= \int_\theta^\infty f_{E_k \mid H_1}(x)\, dx \\ FAR &= \int_\theta^\infty f_{E_k \mid H_0}(x)\, dx. \end{aligned} \tag{5}$$

A graph of P_d versus FAR is called a Receiver Operating Curve (ROC), see [28]. An example for the pdfs in Figure 5a is given in Figure 5b. Observe that it is in fact the choice of the threshold value θ that determines a position on the curve in the graph. What a good choice of P_d and FAR is, is determined by *other* criteria like the cost of causing interference to a primary user (rudeness), or the cost of finding too little unoccupied spectrum (weakness).

Can the detection quality for the k-th frequency bin be improved by combining the hard decisions of N nodes for this bin?

We assume that a particular sensing node receives the decisions of the $N-1$ other nodes through the FOI channel. The question is now how to combine these N local decisions into a global one (aiming at better quality). A possible fusion rule is a rule which decides H_1 if *any* of the N received local decisions is 1 (the

1-out-of-N rule or OR-rule). For this fusion rule the detection probability $P_d^{(N)}$ is [29]:

$$P_d^{(N)} = 1 - (1 - P_d)^N \tag{6}$$

and the $FAR^{(N)}$ is

$$FAR^{(N)} = 1 - (1 - FAR)^N. \tag{7}$$

In Figure 6 the effects of this rule and the number of collaborative nodes are illustrated. By increasing N it can be seen that the ROC curves creep into the upper left corner, signifying improved detection quality.

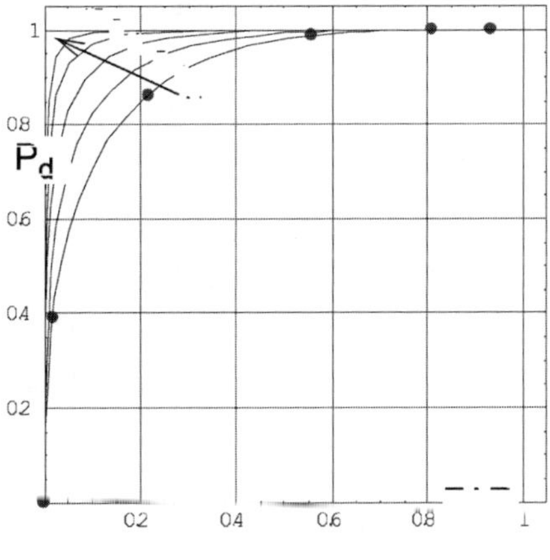

Figure 6. ROC for collaborative scanning (N nodes).

Observe that in this analysis we assumed that all nodes independently "see the same thing (primary user's frequency occupancy)". Apart from the fusion rule shown here, other rules may be applied that result in a different detection quality for the collaborating nodes. In case nodes are in disagreement (e.g. because of fading or shading they do *not* "see the same thing") another analysis is necessary and another fusion rule may outperform the one sketched here.

The approach assumes a more or less errorless instantaneous availability of FOI information for the global decision function (Figure 1). As such the values found by this method of analysis have to be interpreted as a *benchmark* for practical collaborative sensing schemes in which (imperfect) FOI dissemination channels are used.

3. OFDM Based Cognitive Radio Baseband System

In this section we present adaptive OFDM based schemes for Cognitive Radio. We will first discuss adaptive bit loading and adaptive power loading. Second, we will discuss the basic OFDM baseband processing functions in an OFDM receiver.

3.1. Adaptive Bit Loading and Adaptive Power Loading

The basic idea of Orthogonal Frequency Division Multiplexing (OFDM) is to transmit high data rate information by dividing the data into several parallel bit streams where these bit streams are modulated by orthogonal frequencies. The benefit of OFDM is that the high data rate of the whole system is transformed into relatively low data rate streams on each subcarrier which is more robust to inter symbol interference (ISI) caused by multipath delay spread. In the hardware design, it is expected that an OFDM radio is also easy to integrate with spectrum sensing because both of them use FFT cores. The system robustness and hardware resource sharing make OFDM a good choice for Cognitive Radio baseband systems. More importantly, an OFDM system can optimally approach the Shannon capacity in the segmented spectrum by adaptive resource allocation on each subcarrier, which includes adaptive bit loading and adaptive power loading [30]. In an OFDM based Cognitive Radio system, information bits are loaded as different modulation types onto each available subcarrier depending on the subcarrier's signal to noise ratio while the subcarriers currently not available to Cognitive Radio are switched off. Two optimization methods for the adaptive resource allocation can be used for Cognitive Radio: using the power constraint or using the data-rate constraint.

We could maximize the data rate of the system under a certain power constraint. It is formulated as follows:

$$
\begin{aligned}
&\textit{Max}\;\; R = \sum_{k=1}^{K} \frac{F_k}{K} \log_2\Big(1 + \frac{h_k^2 p_k}{N_0 \frac{B}{K}}\Big) \\
&\textit{Subject to:} \sum_{k=1}^{K} p_k \leq P_{total} \\
&F_k \in \{0, 1\} \text{ for all } k \\
&p_k = 0 \text{ for all } k \text{ which satisfies } F_k = 0
\end{aligned}
\qquad (8)
$$

where R is the data rate; K is the number of the subcarriers. N_0 is the noise power density, B is the band of interest for Cognitive Radio, h_k is the subcarrier gain and p_k is the power allocated to the corresponding subcarrier. F_k is the factor indicating the availability of subcarrier k to Cognitive Radio, where $F_k = 1$ means the kth carrier can be used by Cognitive Radio.

The system power minimization can also be applied under the constraint of a constant data rate. We formulate it as follows:

$$\mathit{Min} \sum_{k=1}^{K} p_k = P_{total}$$

$$\mathit{Subject\ to:}\ R = \sum_{k=1}^{K} \frac{F_k}{K} \log_2(1 + \frac{h_k^2 p_k}{N_0 \frac{B}{K}})$$

$$F_k \in \{0, 1\} \text{ for all } k$$
$$p_k = 0 \text{ for all } k \text{ which satisfies } F_k = 0 \qquad (9)$$

A simulation result of adaptive bit loading is shown in Figure 7.

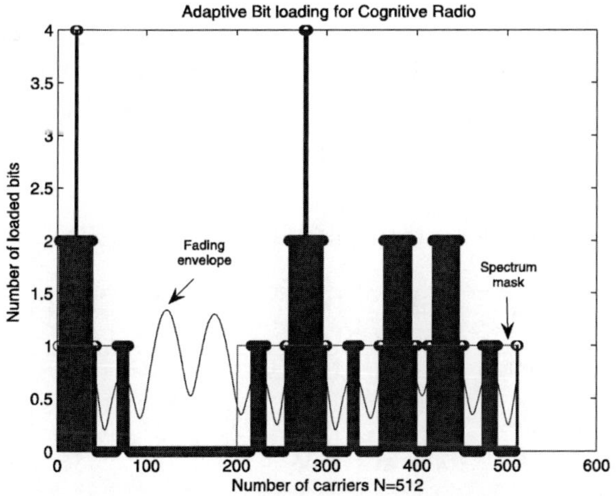

Figure 7. Adaptive bit loading for Cognitive Radio.

In the simulation, we assume a constant data rate of 400 bits for 512 subcarriers in a frequency selective channel. A binary spectrum mask is given by the sensing system to indicate the spectrum availability, a "1" means the subcarrier is available to Cognitive Radio. The optimization method in eq. (9) is applied to achieve constant data rate with power optimization.

3.2. OFDM Baseband Processing

OFDM is widely used in current standards such as HiperLAN/2 [31], DAB [32] and DRM [33]. Due to various system requirements and different radio propagations on different frequency bands, OFDM standards vary a lot in the number of carriers and the transmission bandwidth. Table 3 summarizes the OFDM properties for different standards. Within one standard, multiple modes can be used to cope with various transmission channels by adopting different parameter sets. We take DRM as an example (see Table 4). Mode A is applied for transmissions over Gaussian channels or channels with minor fading which are typical for ground-wave transmissions (medium wave, long wave). The other modes can cope with more time and frequency selective channels which are typical for sky-wave transmissions (short wave and medium wave at night). Mode B and C have the same degree of robustness against delay spread, however mode C has better robustness against Doppler spread because of a shorter useful time T_u. Mode D has the highest robustness against severe delay and Doppler spread among all DRM modes at the cost of the lowest data rate. As we mentioned, Cognitive Radio can operate in different frequency bands, provide multimedia services with various QoS and cope with different channel conditions, therefore Cognitive Radio is able to adapt to different OFDM systems.

Table 3. Properties of the different OFDM standards

	Hiper LAN/2	DAB I	DAB II	DAB III	DAB IV	DRM A	DRM B
Operating frequency[Hz]	5G-6G	174M-240M	1.4G	<3G	1.4G	<30M	<30M
Bandwidth[MHz]	20	1.54	1.54	1.54	1.54	0.012	0.012
Used carrier number	52	1536	384	192	768	226	206
Symbol time [μs]	4	1246	312	156	623	26667	26667
Frame time [ms]	2	96	24	24	48	400	400
FFT size	64	2048	512	256	1024	288	256

Although there are a lot of differences in various OFDM systems, the basic baseband processing is rather similar. This makes OFDM based Cognitive Radio feasible. An OFDM baseband receiver generally consists of the following

Table 4. DRM transmission modes

parameter	mode A	mode B	mode C	mode D
$T_u[ms]$	24	21 1/3	14 2/3	9 1/3
$T_g[ms]$	2 2/3	5 1/3	5 1/3	7 1/3
FFT size	288	256	176	112
used carriers	226	206	138	88

basic tasks (Figure 8):

- **Packet/frame detection and synchronization** is used to determine the starting point of an OFDM frame. This is usually achieved by correlating the received signal with known preambles. We also refer to this frame synchronization as coarse synchronization in contrast with the fine synchronization of the OFDM symbols.

- **Frequency offset estimation and correction** are done to remove the frequency offset which destroys the frequency orthogonality of subcarriers. Frequency offset estimation is done on the frame basis in the preamble section (see Figure 8) before data symbols. A frequency correction coefficient determined by the frequency offset estimation is multiplied by each OFDM data symbol.

- **Channel estimation and equalization** are used to correct the frequency selective fading. Due to the robustness of OFDM to frequency selective fading, less complex frequency domain equalization techniques can be used.

- **Guard time removal** is done after the fine synchronization of OFDM symbols.

- **FFT** is the basic component of all OFDM systems which transforms the signal into the frequency domain.

- **Phase offset tracking and correction** correct the frequency offset residual error by using the pilots in each OFDM symbol.

- **De-map** transforms the complex numbers to a bitstream according to their modulation types. Different modulation types can be loaded on each subcarrier by applying adaptive bit-loading as we have mentioned.

- **De-interleaving** is the opposite task of data interleaving. The idea of data interleaving is that the original data is re-arranged at a transmitter in order to reduce burst errors on a receiver. De-interleaving is the inverse operation of interleaving.

- **Channel decoding** exists in all communication systems not only in OFDM systems. However, OFDM is almost always used in conjunction with channel coding to create coded OFDM (COFDM). A Viterbi code is commonly used, but other channel codes such as turbo code or a LDPC code can also be applied. The code rate can be adaptive to provide different degrees of error protection.

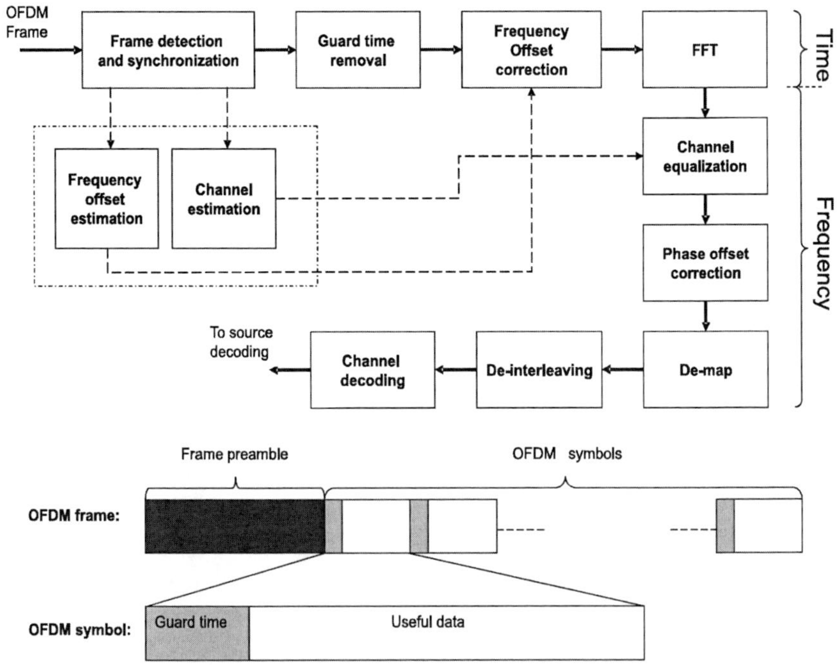

Figure 8. Basic OFDM processing tasks.

For simplicity we only mentioned the tasks performed by the receiver. The transmitter basically does the inverse operations which are less complex. The aforementioned functional blocks can be found in all OFDM systems, but different standards may use different algorithms for each functional block. This means the communication system can select an algorithm to perform each function depending on the requirements of the system. For example, different channel encoders/decoders (codecs) can be applied to achieve different QoS requirements. We refer to this adaptivity as *algorithm-selection level* adaptivity [34]. For a specific algorithm, there are also opportunities for adaptivity by changing parameters of the algorithm. For example the size of FFT and the code rate of Viterbi codec can be tuned by different standards or modes. We refer to this adaptivity as *algorithm-parameter level* adaptivity. The next question is how to achieve the adaptivity for different OFDM systems. A fully reconfigurable hardware platform will be an ideal solution.

4. Heterogeneous Reconfigurable Platform

As already foreseen by Mitola [1], a Cognitive Radio is the final point of software-defined radio platform evolution: a fully reconfigurable radio that changes its communication functions depending on network and/or user demands. His definition on reconfigurability is very broad and we therefore focus on a reconfigurable hardware platform which supports Cognitive Radio. In this section, we present a heterogeneous reconfigurable hardware platform for Cognitive Radio.

4.1. A Heterogeneous Reconfigurable System-on-Chip Architecture

With the evolution of semiconductor technology, more and more transistors can be integrated on a single chip which makes it possible to build large systems, on a chip level rather than on a board level. This approach is called System-on-Chip (SoC). The reconfigurable platform we propose for Cognitive Radio is a heterogeneous reconfigurable SoC architecture shown in Figure 9.

This SoC is a heterogeneous tiled architecture, where tiles can be various processing elements including General Purpose Processors (GPPs), Field Programmable Gate Arrays (FPGAs), Application Specific Integrated Circuits (ASICs) and Domain Specific Reconfigurable Hardware (DSRH) modules. The

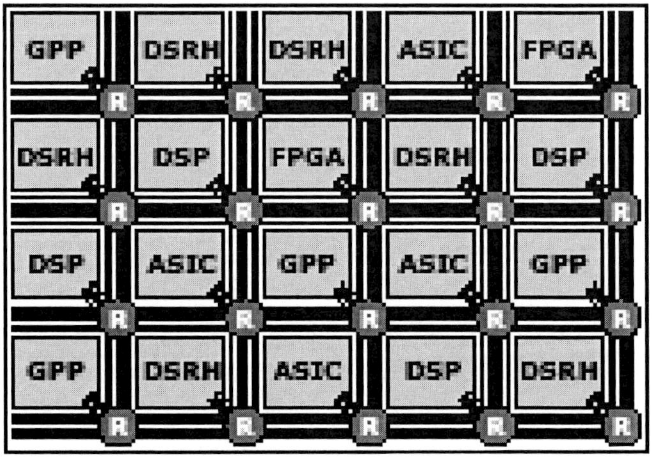

Figure 9. An example of a heterogeneous System on Chip (SoC). DSRH = Domain Specific Reconfigurable Hardware.

tiles in the SoC are interconnected by a Network-on- Chip (NoC). Both the SoC and NoC can be dynamically reconfigurable, which means that the programs (running on the reconfigurable processing elements) as well as the communication links between the processing elements are configured at run-time. Different processing elements are used for different purposes. The general purpose processors are fully programmable to perform different computational tasks, but they are not energy-efficient. The dedicated ASICs are optimized for power and cost. However, they can not be reconfigured to adapt to new applications. FPGAs which are reconfigurable by nature, are good at performing bit-level operations but not that efficient for word level DSP operations. The Domain Specific Reconfigurable Hardware (DSRH) is a relatively new type of processing element, where the configurable hardware is tailored towards a specific application domain. The Montium [35] tile processor (see Figure 10) developed at the University of Twente, and recently commercialised by Recore Systems, is an example of DSRH. It targets the digital signal processing (DSP) algorithm domain, which is the heart of the wireless baseband processing. In our previous work [36] [37] [38], several DSP algorithms used in wireless communication have been mapped onto the Montium architecture. The implementation results show that the Montium architecture is flexible enough to adapt to different algorithms with good energy-efficiency. In a broader sense, working intelligently in an energy-efficient way is an important feature of Cognitive Radio. For Cognitive Radio devices working in the emergency network, energy-efficiency is really a crucial issue because the battery life of radio devices can be a limitation for successful operations. Therefore the reconfigurable platform we propose not only targets flexibility but also energy-efficiency. In the next section, we will introduce the key element of this platform, the Montium tile processor.

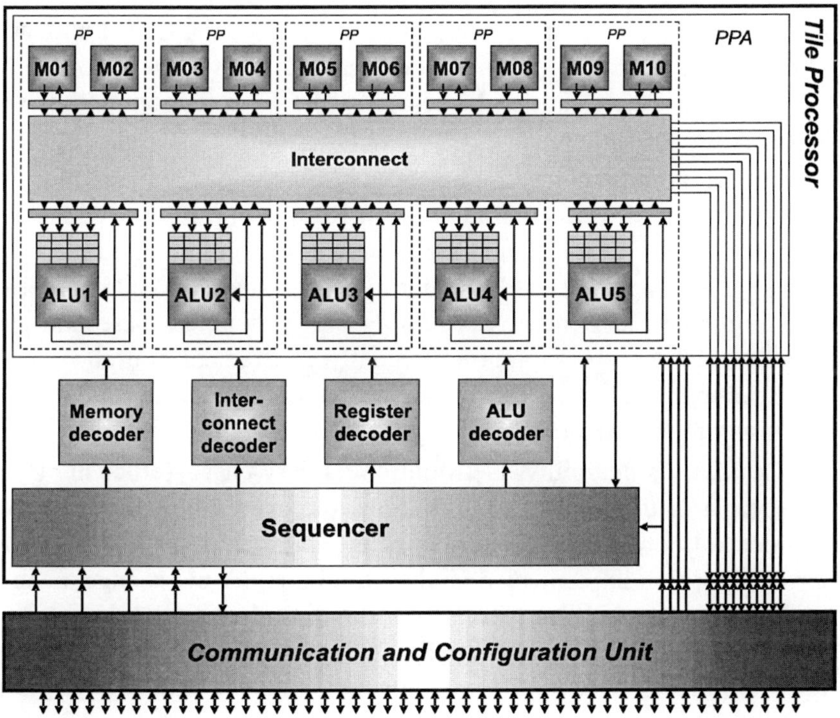

Figure 10. An example of DSRH: a Montium processor.

4.2. The Montium Tile Processor

The Montium is an example of DSRH which targets the 16-bit digital signal processing (DSP) algorithm domain. At first glance the Montium architecture bears a resemblance to a Very Long Instruction Word (VLIW) processor. However, the control structure of the Montium is very different. For (energy-)

efficiency it is imperative to minimize the control overhead. This can be accomplished by statically scheduling instructions and using instruction decoders. The lower part of Figure 10 shows the Communication and Configuration Unit (CCU) and the upper part shows the reconfigurable Tile Processor (TP). The CCU implements the interface for off-tile communication. The definition of the off-tile interface depends on the NoC technology that is used in the SoC. The CCU enables the Montium to run in 'streaming' as well as in 'block' mode. In 'streaming' mode the CCU and the Montium run in parallel (communication and computation overlap in time). In 'block' mode the CCU first reads a block of data, then starts the Montium, and finally after completion of the Montium, the CCU sends the results to the next tile. The TP is the computing part that can be configured to implement a particular algorithm.

The Montium architecture has been refined to perform common DSP algorithms used in wireless communication. Results in [36] showed the Montium's adaptivity to several DSP algorithms including correlation, Finite Impulse Response (FIR) filtering and FFT. Later in [37] an adaptive Viterbi algorithm, considered as the most energy consuming part in wireless baseband processing, was mapped onto the Montium. Recently an irregular non-power-of-two FFT has been mapped [38]. Research is still going on to investigate the Montium's adaptivity to new algorithms.

Besides it's flexibility, the Montium is very energy-efficient. By statically scheduling instructions at compile time, the overhead of both communication and control is reduced. A good example, presented in [35], is that a FIR filtering algorithm implemented on the Montium does not change the instructions in 99% of the time. Therefore, the instruction decoding does not result in excessive switching of control signals. To give a better idea how energy-efficient the Montium is, we show the implementation results of the FFT. This algorithm is one of the most computationally intensive parts, therefore the most energy consuming in OFDM baseband processing. It is common practice to implement these algorithms on dedicated ASICs, however this dedicated hardware can not be reconfigured because, for example, the size of the implemented FFT is fixed. The OFDM based Cognitive Radio we conceive has to be flexible. Therefore a dedicated ASIC is not a solution.

In the OFDM based Cognitive Radio system, an FFT is used by both OFDM baseband processing and spectrum sensing. The size of the FFT has to be adaptive to various OFDM systems. Moreover, spectrum sensing may apply coarse sensing or fine sensing with variable sized FFTs. Let us consider the most com-

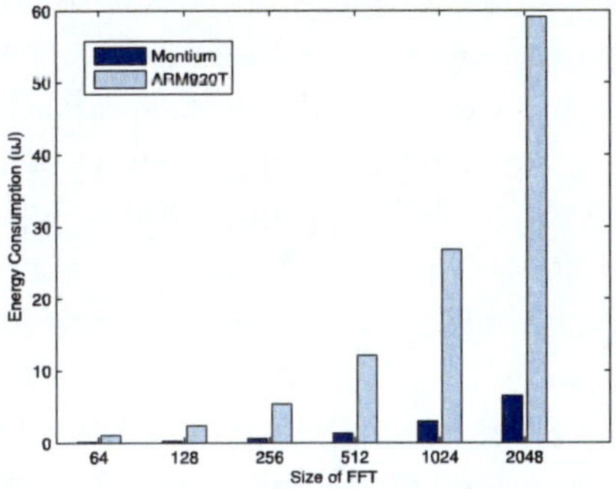

Figure 11. Energy consumption (in μJ) comparison between the Montium and ARM.

mon radix-2 FFT. As the size of FFT increases, the computational complexity of the FFT increases on a logarithmic scale because the order of the complexity is $\frac{N}{2}\log_2 N$. So, FFT processing especially for large size FFTs have tough demands for the processor in terms of timing and energy consumption. A general purpose embedded solution like the ARM (a reduced instruction set computer (RISC) processor architecture) is inefficient for these type of applications. According to [35], the execution part of an FFT butterfly takes 21 clock cycles on an ARM920T running at 250MHz while it takes only 1 clock cycle on the Montium running only at 100MHz. Put in a simple way, a Montium is about ten times faster than an ARM. The energy consumption of the Montium is significantly lower than of the ARM, as illustrated in Figure 11. For fair comparison, the ARM implementation is optimized and both implementations are implemented in 0.13 μm technology. If a 1024 points FFT is being executed for spectrum sensing while another 1024 points FFT is required for OFDM baseband processing at the same time, the energy saving of the Montium implementation will be significant compared with the ARM implementation as shown in Figure 11.

4.3. The Design Methodology

To implement Cognitive Radio on a heterogenous reconfigurable platform, we propose a design methodology which has two new features: 1) Transaction level modelling of an application into a parallel task graph, 2) Run-time spatial mapping of tasks onto heterogenous processing tiles.

First, a high level application such as the baseband processing part of a Cognitive Radio, is partitioned into tasks. The applications are typically streaming applications. The tasks within the applications communicate with each other via channels which are accessed by tasks through interfaces. Communications between tasks are data transactions including synchronization and data transfer. In a transaction level model, the details of communication among computation components are separated from the details of the computation components. For example an OFDM receiver can be modelled by the computation components (or tasks) we have mentioned in the previous chapter while communications are modelled at transaction level. In our implementation model, we use the Task Transaction Level (TTL) approach [39]. Here we give a pseudo code example of the FFT task implementation.

```
process Task_FFT {
 initialization;
 while (true)
 {\\ declaration of local variables
  complex Y[64],Z[64];
  for(i=0;i<64; i++)
  \\ read input samples
  InRead(&(Task_FFT->inport), &Y[i]);
  \\ perform the FFT
  call function Z=FFT(Y);
  for(i=0; i<64; i++)
  \\ write output samples
  OutWrite(&(Task_FFT->outport), &Z[i]);
  }
}
```

In principle, the FFT size is application dependent but in the example we assume the FFT size is 64. While the condition is true, 64 samples are read into the local buffer from the channel which is connected to the input port. When 64 samples are ready, data processing is executed by calling the FFT function. This FFT function can be implemented on various processing tiles like ARM, FPGA or Montium. Results are written to the channel connected to the output port. Both synchronization and data transfer are done by simple read and write function calls. From the example we can see that the computation components can be plugged-in and replaced as functions, which allows application adaptivity within the same TTL framework. So, it is advantageous to use the transaction level model to speed up simulation and allow exploring and validating design alternatives at a high level of abstraction.

For each task of an application, one or more task implementations have to be provided. A task implementation is the implementation of a task on a particular tile, e.g. object code for an ARM or a DSP or configuration data for an FPGA or a Montium. A task implementation has several characteristics, e.g. the amount of energy it takes to execute the task on a particular tile of the architecture, the delay or tile utilization. This implementation library (including the characteristics) is composed at design-time.

At run-time, a spatial mapping function module selects the tile implementation for each task of an application. The objective of mapping is to find a mapping solution which optimizes a certain design metric such as the energy consumption given other system constraints (e.g. timing, tile availability). Such a mapping problem is known to be NP-hard and in practice, the corresponding optimization algorithms may run for hours. Therefore, heuristics have to be used to find a solution with a reasonable quality within an acceptable time. In [40], a minimal weight algorithm is proposed to map a digital audio broadcast receiver onto a tiled SoC in a reasonable time, where the optimization metric is to minimize the energy consumption. Run-time mapping offers a number of advantages over design-time mapping, especially for a Cognitive Radio which dynamically changes its behavior.

- First, run-time mapping can adapt to the available resources. The available resources may vary over time due to multiple applications running simultaneously.

- Second, run-time mapping can adapt to dynamically changing applications. A task within the application may change its computational behavior at run-time, for example, the size of the FFT or a new task can be added to the application.

5. Conclusion

In this chapter, we give an introduction into Cognitive Radio in the context of emergency networks. Our focus on Cognitive Radio is to search for under-utilised spectrum and to (rapidly) adapt transmission settings accordingly to meet various system requirements of emergency networks. Therefore, the research on Cognitive Radio concentrates on frequency scanning and adaptive baseband processing. Three architectures for single-node energy-based scanning have been identified: scan-based, real-time FFT based and scanning-FFT based. The scanning-FFT based approach is shown to be most promising. By introducing collaboration between sensing nodes, an increase in detection quality can be expected.

The robustness to multipath delay spread and the hardware resource sharing with scanning FFT-based spectrum sensing, make OFDM a good choice for the Cognitive Radio baseband system. Moreover, OFDM systems that can nullify individual carriers poses interesting opportunities for Cognitive Radio. A highly adaptive OFDM system is foreseen since it has to operate in different frequency bands, it has to deal with various effects of wireless channels like fading and shadowing and it has to support various multimedia services. In order to support a flexible Cognitive Radio, a heterogenous reconfigurable SoC platform is required, similar to a Software Defined Radio. The key element in the platform, the coarse-grain reconfigurable tile processor (the Montium), enables the reconfigurability in combination with the energy-efficiency. A design methodology is needed to map applications onto a heterogeneous platform which has two new features: transaction level modelling of applications and run-time spatial mapping.

References

[1] Mitola, J. III. *Cognitive Radio: An Integrated Agent Architecture for Software Defined Radio.* PhD thesis, Royal Institute of Technology, Sweden, May. 2000.

[2] Haykin, S. "Cognitive Radio: Brain-empowered Wireless Communication". *IEEE J. Select. Areas Commun.*, vol. 23, no.2:pp 201–220, Feb. 2005.

[3] http://www.freeband.nl/project.cfm?id=488.

[4] McHenry, M. A. "NSF spectrum occupancy meassurements project summary". http://www.sharespectrum.com.

[5] Tewfik L. Doumi. "Spectrum Considerations for Public Safety in the United States". *IEEE Commun. Mag.*, Jan. 2006.

[6] Weiss, T.A. and Jondral, F.K. "Spectrum Pooling: An Innovative Strategy for the Enhancement of Spectrum Efficiency". *IEEE Commun. Mag.*, vol. 24, no.3:pp S8–S14, Mar. 2004.

[7] Mitola, J. "Cognitive Radio: Making software radios more personal". *IEEE Pers. Commun.*, Aug. 1999.

[8] Federal Communication Commision. "Spectrum Policy Task Force". Tech. Report., Nov. 2002.

[9] Robert W. Brodersen, Adam Wolisz, Danijela Cabric, Shridhar Mubaraq Mishra, Daniel Willkomm. "A Cognitive Radio Approach for Usage of Virtual Unlicensed Spectrum". White Paper, July 2004. `http://bwrc.eecs.berkeley.edu/Research/MCMA/CR_White_paper_final1.pdf`.

[10] Shridhar Mubaraq Mishra et al. "A real time Cognitive Radio testded for physical and link layer experiments". In *Proc. of the IEEE Symposium on New Frontiers in Dynamic Spectrum Access*, pages 562–567, Nov. 2005.

[11] http://bwrc.eecs.berkeley.edu/Research/Cognitive.

[12] Weiss, T., Krohn, A., Hillenbrand, J. and Jondral, F. "Mutual Interference in OFDM-based Spectrum Pooling Systems". In *Proc of VTC*, Milan, Italy, May 2004.

[13] http://www.cwt.vt.edu/research/cognitiveradio/index.html.

[14] Christian James Rieser. *Biologically Inspired Cognitive Radio Engine Model Utilizing Distributed Genetic Algorithms for Secure and Robust Wireless Communications and Networking*. PhD thesis, Virginia Tech.

[15] http://e2r.motlabs.com.

[16] http://www.ieee802.org/22.

[17] Cabric, D., Mishra, S.M. and Brodersen, R.W. "Implementation Issues in Spectrum Sensing for Cognitive Radios". 38th Asilomar conference on signals, systems and computers, pages 772–776, 2004.

[18] Pelgrom, M. "Analog-to-digital Conversion", March 2003. Course Notes University of Twente, course 121048.

[19] Lundberg, K.H. "High-Speed Analog-to-Digital Converter Survey", October 2002. Accessible via `http://web.mit.edu/klund/www/papers/`.

[20] Sanders, F.H. and Lawrence, V.S. "Broadband Spectrum Survey at Denver, Colorado". NTIA Report TR-95-321, NTIA, September 1995. `http://www.its.bldrdoc.gov/pub/surv_dnv/`.

[21] McHenry, M. and McCloskey, D. "New York City Spectrum Occupancy measurements, September 2004". Report, Shared Spectrum Company, December 2004.

[22] Agilent. "Spectrum Analyser Basics". Application Note 150, Agilent, 2004.

[23] Hoeksema, F.W., Heskamp, M. , Schiphorst, R. and Slump, C.H. "A Node Architecture for Disaster Relief Networking". 1st IEEE International Symposium on new Frontiers in Dynamic Spectrum Access Networks(DySPAN'05), pages 577–584, November 2005.

[24] Agilent. "Swept and FFT Analysis". Performance spectrum analyzer series - application note, Agilent, 2004.

[25] Tektronics. "Fundamentals of Real-Time Spectrum Analysis". Technical report, *Tektronics*, 2004.

[26] Radiocommunications Agency The Netherlands) Dutch Ministry of Economic Affairs (Agentschap Telecom. "Nationaal Frequentieplan 2005 (National Frequency Plan 2005)". `http://www.ez.nl/content.jsp?objectid=24642`, February 2005.

[27] Radiocommunications Agency The Netherlands) Dutch Ministry of Economic Affairs (Agentschap Telecom. "Nationaal Frequentieregister (National Frequency Register)". `http://www.agentschap-telecom.nl/nfr/`, March 2005.

[28] van Trees, H.L. *Detection, Estimation, and Modulation Theory*, volume I: Detection, estimation, and Linear Modulation Theory. John Wiley & Sons, 1968.

[29] Varshney, P.K. *Distributed Detection and Data Fusion.* Springer-Verlag, 1997.

[30] Budiarjo, I., Nikookar, H., and Ligthart, L.P. "Overview of Adaptive OFDM in the Context of Cognitive Radio". In *12'th IEEE Symposium on Communications and Vehicular Technology in the BENELUX*, Enschede, the Netherlands, Nov. 2005.

[31] ETSI. "Broadband Radio Access Networks (BRAN); HIPERLAN Type 2; Physical (PHY) layer". Technical Specification ETSI TS 101 475 V1.2.2 (2001-02), 2001.

[32] ETSI. "Radio broadcasting systems: Digital Audio Broadcasting to mobile, portable and fixed receivers". Technical Specification EN 300 401.

[33] ETSI. "Digital Radio Mondiale (DRM) System Specification". Technical Specification ETSI TS 101 980 V1.1.1 (2001-09), 2001.

[34] Gerard K. Rauwerda, Jordy Potman, Fokke W. Hoeksema, and Gerard J.M. Smit. "Adaptation in the physical layer using heterogeneous reconfigurable hardware". In *Adaptation Techniques in Wireless Multimedia Networks*, 2006. ISBN 1-59454-883-8.

[35] Paul M. Heysters. *Coarse-Grained Reconfigurable Processors; Flexibility meets Efficiency*. PhD thesis, University of Twente, September 2004. ISBN 90-365-2076-2.

[36] Paul M. Heysters and Gerard J.M. Smit. "Mapping of DSP Algorithms on the Montium Architecture". In *Proceedings of Reconfigurable Achitecture Workshop*, 2003.

[37] Gerard K. Rauwerda, Gerard J.M. Smit, and Werner Brugger. "Implementing an Adaptive Viterbi Algorithm in Coarse-Grained Reconfigurable Hardware". In *Proceedings of the International Conference on Engineering of Reconfigurable Systems and Algorithms*, 2005.

[38] Arnaud Rivaton, Jérôme Quevremont, Qiwei Zhang, Pascal T. Wolkotte, and Gerard J.M. Smit. "Implementing Non Power-of-Two FFTs on Coarse-Grain Reconfigurable Architectures". In *Proceedings of the International Symposium on System-on-Chip (SoC 2005)*, 2005.

[39] Pieter van der Wolf et al. "Design and Programming of Embedded Multiprocessors: An Interface-Centric Approach". In *Proceedings of the CODES+ISSS*, 2004.

[40] Smit, L.T., Hurink, J.L. and Smit. G.J.M. "Run-time mapping of applications to a heterogeneous SoC". In *IEEE 2005 International Symposium on System-on-Chip Proceedings*, Tampere, Finland, 2005.

Errata for *Cognitive Radio Terminology, Technology and Techniques*

Editors: Peyton D. McGuire and Harry M. Estrada; **ISBN**: 978-1-60876-604-8

p. 237

The chapter number should be 8.

In: Cognitive Radio
ISBN 978-1-60876-604-8
Editors: P.D. McGuire et al, pp. 237-262

Chapter 9

COGNITIVE RADIO WIRELESS NETWORKS

***Siva Subramani*[*], *Zhong Fan*[2†], *Simon Armour*[1‡] and *Dritan Kaleshi*[1§]**

[1]Centre for Communications Research, University of Bristol, UK
[2]Toshiba Research Europe Limited, Bristol, UK

Abstract

Cognitive radios are intelligent wireless systems that learn about their environment and adaptively optimize their performance by modifying their transmission parameters. The promising prospects of cognitive radio to be an advancement in the field of wireless networking and a solution for the spectrum under-utilization problem has made it an independent research discipline on its own. The dynamic and adaptive nature of the radio has many challenges in terms of practical realization. In this chapter, an overview of the cognitive radio technology and the latest research developments in different layers of the protocol stack are given. The important topics such as wideband spectrum sensing, transceiver architecture and dynamic spectrum allocation are discussed in detail.

PACS 05.45-a, 52.35.Mw, 96.50.Fm.

Keywords: Cognitive radio.

[*]E-mail address: Siva.Subramani@bristol.ac.uk
[†]E-mail address: Zhong.Fan@toshiba-trel.com
[‡]E-mail address: Simon.Armour@bristol.ac.uk
[§]E-mail address: Dritan.Kaleshi@bristol.ac.uk

Key Words: Reconfigurable radio, spectrum sensing, dynamic spectrum access.

AMS Subject Classification: 53D, 37C, 65P.

1. Introduction

In recent years the advancements in wireless devices and technologies have seen an unprecedented growth and this trend is envisaged to continue for years to come. With the advent of wireless broadband and futuristic home network concepts, this proliferation of wireless devices will see a dramatic usage in everyone's day to day life. Moreover, the wireless applications would include not just simple data or voice but high data rate video and realtime multimedia links. Due to the currently followed scheme of long-term licensing of radio frequency spectrum and inflexible regulations in access technologies, the communication traffic is projected to grow only in the selected portions of spectrum. Recent empirical measurements show that the vast majority of the radio spectrum is underutilized [1] [2]. The spectrum utilization depends on the geographical location and time of measurement. Independent studies carried out in Berkeley, California and Heathrow airport, UK shows the average utilization of 6%. The utilization pattern shows some bands (including the bands in mobile communication, unlicensed ISM bands[1]) are in high demand and utilization. There are other bands of spectrum which are rarely used, subject to geographical location and time. This high utilization and demand for certain portions of spectrum and under-utilization of the remaining parts sparked the need for immediate reforms and for flexible spectrum policies [3]. Spectrum regulatory agencies such as FCC (USA), Ofcom (UK) and others in respective countries are developing strategies to improve spectrum utilization and occupancy patterns [4]. In recent years, spectrum regulators and researchers are seriously considering short-term leasing or allowing unused spectrum bands to be shared. Relaxing the spectrum access policies, regulations and allowing spectrum sharing could significantly improve utilization.

Cognitive radios are autonomous reconfigurable radios that learn about their environment and adaptively optimize their performance by modifying their transmission parameters, including switching transmission channels. The exceptional developments in the field of reconfigurable and software defined radio

[1]Industrial, Scientific and Medical (ISM) are non-commercial bands; bands in 900MHz, 2.4GHz and 5.8GHz being the significant ones.

(SDR) and the evolution of related technologies such as signal processing, interference mitigation, radio resource management (RRM) which addresses application QoS requirements and effectively exploits diversity and adaptive access algorithms make cognitive radio an ideal technical candidate solution for the use of underutilized and shared spectrum access. The reconfigurable software defined radio technology that has evolved into cognitive radio with the attributes of learning and adaptivity has the following desirable features:

- Reconfigurable - tunable to any frequency and transmit and receive with any modulation
- Awareness - RF environment aware, location aware
- Intelligent in decision making and adaptive in operations
- Unambiguous communication and the ability to coexist with other types of radios (legacy systems)

In this chapter, the network formed by these cognitive radios coexisting with the other radios is referred to as a cognitive radio network. In the context of cognitive radio networks, the cognitive radios may be referred to as Secondary Users (SU) and the licensed users as Primary Users (PU). Cognitive radios are opportunistic radios that sense and access the unoccupied spectrum and evacuate the channel on the arrival of primary users without causing interference.

1.1. Challenges and Motivation

Cognitive Radio (CR) is a visionary field transforming into a research discipline on its own. But realizing cognitive radio is a challenge. The promising prospects of cognitive radio to be an advancement in the field of wireless networking and solution for the spectrum under-utilization problem has raised many expectations. Cognitive radio has become an independent research discipline and the technical areas span from the PHY layer, signal processing, information theory to medium access, coexistence and RF front-end enabling components. The successful development of a cognitive radio network could result in improved spectrum utilization and a big leap in technological advancement in the field of communication and networks.

Cognitive Radio has been an active area of research in the field of communications and networks in the last decade, rigorously intensifying in the last three to four years. In this chapter, the challenges and the latest research developments in each of the research disciplines are described.

If regulatory agencies open up the spectrum for sharing, then the spectrum that would be available for secondary users will likely be non-contiguous frequency bands of different width, distributed over a wide range of spectrum. The wide-band spectrum sensing and accurate detection of signals is still a challenging problem. Autonomous sensing and how cooperative sensing could improve the detection process is an active area of research. Additionally, cognitive radio solutions will require the ability of reliably scanning the spectrum to find a free channel. Wide band scanning and the receiver architecture that enables this wide band sensing is an interesting research theme. Once the spectrum is sensed to be available the framework of dynamic access mechanism studies the methods to access the channel and coexist with the other similar cognitive radios. The channels are either allocated by a central entity (eg. base station or access point) or autonomously accessed as in listen-before-talk (LBT) methods.

The remainder of the chapter is organized as follows: Section 2 describes various sensing techniques and the methods to improve the detection probabilities. The later part of this section discusses the wide band scanning and estimation issues. Section 3 explains the enabling receiver architecture, particularly relating to wide band sensing and hardware limitations. Section 4 elaborates on the spectrum allocation methods, considering both the centralized and distributed type networks. Section 5 concludes with notes on other research themes and references.

2. Spectrum Sensing and Detection

One of the fundamental working principles of cognitive radio is based on the effectiveness of sensing and estimating the available spectrum bands. The spectrum usage of some of the licensed (primary) users is sporadic and exhibits dynamic nature. Accurate and effective sensing of a wide-band spectrum is a challenging problem, particularly in distributed cognitive radio networks.

2.1. Cognitive Radio Sensing Methods

In this section, some of the important sensing techniques are described before looking into improving the detection performance. An overview of some of the well-known sensing techniques is given. The following sensing techniques could be used in stand-alone radios in a distributed type network. In the later section, cooperation or collaboration of distributed cognitive radios to improve detection is discussed.

The essence of spectrum sensing is a binary hypothesis-testing problem:

$$\mathcal{H}_0 : \text{Primary user is absent}$$
$$\mathcal{H}_1 : \text{Primary user is present in that channel}$$

2.1.1. Energy Detection or Radiometry

An energy detector is one of the simplest and most common methods of sensing a channel. This technique does not need prior information on the band or signal type since the detector makes a decision on the above binary hypothesis based on the power of the signal present in the spectrum band under consideration. Consider the received signal with the following form:

$$y(n) = s(n) + w(n) \tag{1}$$

where $s(n)$ is the signal to be detected, $w(n)$ is the Additive Gaussian White Noise (AWGN) sample, modeled as a zero-mean random variable with variance σ^2, i.e. $w(n) = \mathcal{N}(0, \sigma^2)$. The normalized power in the received signal is given by, $T(y) = \frac{1}{N}\sum_{n=1}^{N} |y[n]|^2$. The decision rule then compares the energy estimate of the observed sequence with a threshold, γ.

$$\begin{cases} \mathcal{H}_1 & \text{if} \quad T(x) \geq \gamma \\ \mathcal{H}_0 & \text{otherwise} \end{cases}$$

The performance of the detector can be summarized with two probabilities: probability of detection, P_D, and probability of false alarm, P_F. P_D is the probability of detecting a signal on the considered channel when the signal is truly present. It can be shown that as,

$$P_D = Pr(T > \gamma|\mathcal{H}_1) \tag{2}$$

P_F is the probability that the test incorrectly decides that the channel is occupied when it is not. The performance of the detector is based on the selected threshold. Since the performance is based on the power detected, the energy detector cannot differentiate between PU and SU signals. Moreover, the performance is limited for detecting low SNR (Signal to Noise Ratio) signals due to the uncertainties in noise power. The limitations and the potential solutions are discussed later in this section.

In [5] R. Tandra *et al.* derive analytical bounds on detection performance and explain limitations of low SNR signals under Gaussian noise. These bounds and analysis are used to understand the detector and need for training data or pilot tone for primary signals. In [6] the authors show that presence of a pilot signal could improve the performance of the detector.

2.1.2. Matched Filter Detection

For deterministic signals, matched filter detection is considered to be the optimal signal detection since it maximizes signal to noise ratio. A matched filter detection works by correlating a known signal, or template, with an unknown signal to detect the presence of the template in the unknown signal. The detector has to effectively demodulate the signal during detection, which means the radio should be aware of the modulation type and order, codebook, etc., For cognitive radios, the detection and demodulation is still possible because the primary signals (e.g., radar) typically use pilot signals, preambles or carrier synchronization. However, the main downfall is that the matched filter method needs dedicated receivers for different classes of primary signals [7].

2.1.3. Cyclostationary Detection

Cyclostationary detection is a method for detecting primary user signals by exploiting the cyclostationary features of the received signals [8]. A signal is said to be cyclostationary (in the wide sense) if its autocorrelation is a periodic function of time t with some period. The Cyclic Autocorrelation Function (CAF) $R_y^{\alpha}(\tau)$ of the observed signal $y(t)$ is given by,

$$R_y^{\alpha}(\tau) = E[y(t+\tau)y^*(t-\tau)e^{j2\pi\alpha t}] \tag{3}$$

where α is the cyclic frequency. Then the Cyclic Spectral Density (CSD) can be calculated as

$$S(f,\alpha) = \int_{\infty}^{\infty} R_y^{\alpha}(\tau)e^{-j2\pi f\tau} \tag{4}$$

The CSD function gives the peak value when the cyclic frequency is equal to the fundamental frequencies of the transmitted signal $x(t)$. This distinctive character of spectral redundancy makes signal selectivity possible. Different types

Table 1. Comparison of Spectrum Sensing Methods

Detector	Advantages	Disadvantages
Energy detector	Simple architecture	Not efficient for low SNR signals
	Do not need any prior information	
Matched filter detector	Optimal detector	Need prior information
	Relatively simpler architecture	
Cyclostationary detector	Efficient for low SNR	Computationally complex
	signals	Works for periodic signals
Wavelet detection	Efficient for wide band	Computationally complex
		Ineffecient for spread spectrum signals

of modulated signals such as BPSK and QPSK that have identical power spectral density functions can have high spectral correlation functions. Whereas, stationary noise or interference exhibits no such correlation functions.

2.1.4. Wavelet Detection

Sensing wideband spectrum with multiple sub-bands can be viewed as an edge detection problem. In this technique, wideband spectrum is considered as the continuous sub-bands where the power spectral characteristics are smooth within bands but changes abruptly on the border of bands. By employing a wavelet transform of the power spectral density (PSD) of the observed signal $y(t)$, the singularities of the PSD $S_y(f)$ can be located and thus the vacant frequency bands can be found. The wavelet detection approach is particularly useful when detecting non-contiguous bands in the wide spectrum. In [9] Tian and Giannakis proposed a wavelet approach to wideband spectrum sensing for cognitive radios. The solution was based on the gradient wavelet modulus and multiscale wavelet products.

Table 1 provides a comparison of different sensing methods.

2.2. Cooperative Sensing

The performance of a stand-alone detector is limited due to the geographical location or limited information for confirming detection hypothesis. Some of the common factors affecting the detection performance are fading, shadowing and noise uncertainties. For example, when experiencing a deep fade cognitive radio cannot reliably sense the presence of the primary user due to the low SNR of the received signal. Then the cognitive radio incorrectly observes the channel as vacant and begins accessing the channel. This common problem is referred to as hidden terminal problem. The probability of detection could be improved through cooperative sensing. There are several cooperation sensing schemes where cognitive users cooperate or collaborate to mitigate false detection and improve spectrum sensing in shadowing and fading environments [10][11][12]. The cooperation schemes reduce the mis-detections and the probability of false alarms considerably.

The cooperation or collaboration in cognitive radio networks could be categorized into two classes based on the network architecture, centralized networks and distributed networks.

2.2.1. Cooperative Sensing in Centralized Networks

In this section a network of cognitive radios with a centralized entity (a base station (BS) or access point (AP) or a leader node) is considered. The three fundamental cooperative sensing tasks are (i) local sensing, (ii) data collection, and fusion (iii) decision making. The cognitive radios sense the band of frequencies of interest locally and the information is transmitted to the centralized entity. It is worth noting that data collection and fusion is computationally complex and involves a lot of power. Potentially, the exchange of channel sensing information can take place over a common control channel. In the case of leader node, the selection of leader should take this into account. Data fusion involves convolution of the collected sensing data. With the resulting information a binary-hypothesis decision on the frequencies and users is formed. As a typical example, the secluded shadowed or faded cognitive radio will be able to form a decision with this cooperation.

2.2.2. Cooperative Sensing in Distributed Networks

In the absence of a centralized entity, the cognitive radios form a distributed type (ad hoc) network. The three fundamental tasks remain the same but are adopted to the distributed network. The sensing information can be shared among the cognitive radios to improve the probability of detection (or reduce probability of false alarm). Cooperative sensing in a distributed network is a challenging problem. For coordination among the nodes, trust (reliability) plays an important role. There are several interesting cooperative sensing algorithms recently proposed to address these issues. The two steps of the cognitive cycle, observation and analysis, are done in the stand alone cognitive radios. The sensing information is then shared among the nodes. And if a cognitive radio requires additional sensing information for access or confirmation it sends requests to nodes in the network.

In [14], a distributed collaborative sensing algorithm GUESS (Gossiping Updates for Efficient Spectrum Sensing) was proposed which has fast convergence and reduced protocol overhead and is robust to network changes. In [15], the authors proposed a distributed spectrum sensing technique. In this paper, time frequency analysis is used to classify the air-interfaces. The bilinear nature of the Time Frequency (TF) transforms provides a methodology to process time-varying and superimposed signals. The features are extracted through the Karhunen-Loeve (KL) method. The KL method is a classical eigenvalue method for linear feature reduction where the optimal weights for features combinations are computed on the basis of contributions of features with training set. In [12], the authors showed the cooperation could reduce the detection time of the "weaker" node. The cooperative scheme reduce the overall detection time and hence improving agility in the network. Moreover, the cognitive radios could take advantage of the adaptive nature and vary the power. In a fading environment, the cognitive radio could opportunistically transmit at high power levels. The capacity gains during such opportunistic spectrum access in the fading channel was analyzed in [16].

Furthermore, the performance of cooperative sensing increases with the number of users in the networks. With a greater number of users the sensing information is more accurate and the probability of false alarm is drastically reduced. However, the cooperation schemes have some bottlenecks if not issues. The cooperative sensing depends on the common channel and the capacity of the common channel. With a large number of cognitive radios, the total number of sensing bits transmitted to the common receiver tends to be very large and this will require high control channel bandwidth. The censored decision approach is used to exclude an ambiguous detection region [17]. By setting the detection interval $[\lambda_1, \lambda_2]$, only the radios having observation values out of this region are required to report to the common receiver.

2.3. Spectral Scanning and Estimation

Accurate and effective sensing of a wide-band spectrum is a challenging problem in distributed cognitive radio networks. Assessing the dynamic nature of the spectrum and gathering reliable and sufficient information is essential for opportunistic spectrum access. In this section the statistical spectrum estimation methods are reviewed with a note on the inefficiencies in analyzing the space-time statistics. A new wideband spectrum scanning method is also discussed.

2.3.1. Statistical Spectrum Estimation

The stimuli generated by wireless radios are *nonstationary spatio-temporal* signals in that their statistics depends on both time and space. The statistical analysis of the nonstationary signals is accounted for better by adopting *time-frequency distribution* of the signal. The traditional periodogram method of spectral estimation using FFT (Fast Fourier Transform) is a badly biased and inconsistent estimator of the power spectrum, opinioned in [18], since this method assumes signals to be deterministic and matches marginal distribution. The stochastic approach to the time-frequency distribution under the Loève transform is widely advocated. However, for wireless communications, if the continuous RF stimuli exhibits pseudostationarity then multitaper spectral estimation is considered to be the method of choice. The biasing of the power-spectrum estimate due to sidelobe leakage and loss of information is reduced by multiple *tapering* (i.e., windowing). More detailed studies on multitaper analysis of nonstationary signals are given by Thomson in [19] [20].

In [21] Harris has presented optimal spectral estimator for cognitive radios by considering window selection, window length, window overlap and ensemble average options. For example, the difficulty of making spectral measurement of noise like signals due to poor statistics of the sum of square terms obtained from the output of the transform has motivated to ensemble moving averages and exponential averages. In [22] Rojas *et al.* considered a filter-bank framework where the shape and bandwidth of the candidate spectral component instead of a single carrier.

2.3.2. Wideband Spectrum Scanning Methods

A scanning mechanism controls the sequence of scanning and the portion of spectrum to be scanned and it works in conjunction with the sensing mechanism. In a dynamic spectrum sharing network, when spectrum occupancy is non-uniformly distributed, scanning multiple spectrum regions simultaneously could provide diverse and sufficient spectrum information. Conventional sequential scanning lacks the ability to find a free-channel in multiple spectrum region and may not be effective. For effective sensing over the spectrum, a bidirectional and dual scanning mechanism is analyzed in [23] to search over a broader and multiple spectrum space.

A cognitive radio regularly searches for a free-channel when one of the following events happens

1. Initial channel selection: When a SU scans the spectrum for free channel for initial access. or,

2. Unexpected channel switching: When a SU evacuates the current channel and switches to another channel due to a primary user arrival or when the channel quality suddenly degrades

2.4. Bidirectional Scanning Method

In sequential unidirectional scanning, the radio scanner scans the channel in steps either in incrementing or decrementing order. Since the search space is limited to one direction, a unidirectional scan could miss a free channel available in the other direction and therefore does not ensure the detection of the nearest available free channel. By adopting bidirectional scanning, the scanning is performed in both directions, expanding through the spectrum space in both directions finding the nearest available channel.

Figure 1 shows the current band (band X) and the scanning methodology. With the starting frequency f_x and the center frequency of channel c_x, the bidirectional scanner scans channels in either direction. The scanning steps in either direction are selected alternatively and expand in either side as c_{x+1}, c_{x-1}, c_{x+2}, c_{x-2},... and so on, as shown in Figure 1. This alternate bidirectional scanning steps appears to be scanning in a *spiral* fashion and hence may be referred to as spiral-bidirectional scanning.

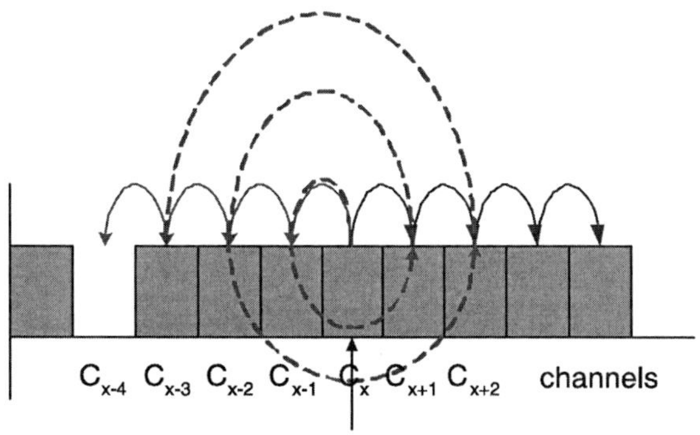

Figure 1. Bidirectional Spectrum Scanning method.

2.5. Dual Scanning

In dual scanning, the scanner initiates scanning at two different spectrum bands, as shown in Figure 2. The starting points can be pseudo randomly selected or selected in such a way that one is the current channel and the other is sufficiently well spaced from it within the available range. This dual scanning method combines random and sequential scanning, now capable of reaching multiple spectrum bands in a controlled manner. The spectrum bands are alternatively checked. If a free-channel is found matching the requirement, the information is mutually shared and agreed for spectrum access. The boundary conditions ensure that the same band is not scanned twice within the revisit time.

If one of the search coordinates reaches the boundary of the allowed spectrum, the search switches to unidirectional from the other edge and continues searching. This method is called Bidirectional + Unidirectional. When the spectrum occupancy is non-uniform and the current scanning region is densely occupied, then switching the scanning sequence to a distant spectrum region away from the current scanning area is an alternate option. This method is referred to as Bidirectional + Random, due to the relative random jump to a different spectrum band.

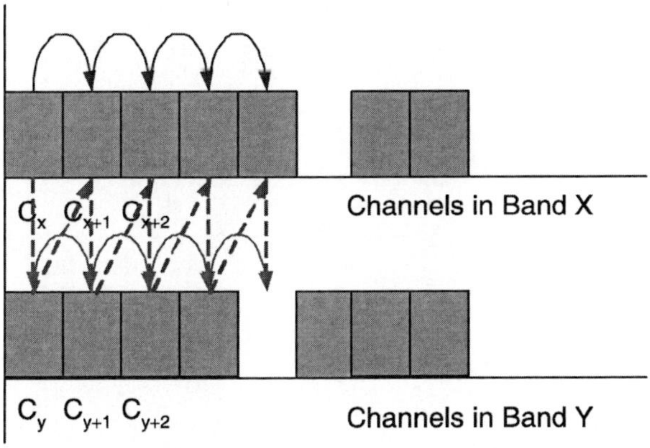

Figure 2. Dual Spectrum Scanning method.

3. Enabling Architecture and Components

Sensing the spectrum is an important functionality of the cognitive radio receiver. The sensing functionality could range from finding a free-channel, detecting any primary user's arrival to estimating the suitability of a channel. There are several types of receiver models available.

1. Receiver incorporating sensing mechanism
2. Simple but independent sensing mechanism in addition to the receiver
3. Intelligent and independent sensing mechanism in addition to the receiver

Each model has its advantages and disadvantages depending on the battery power consumption, cost of additional hardware, algorithm complexity, etc. Incorporating sensing functionality to the existing radio front-end architecture could impose additional complexity. Moreover, the receiver would not be able to sense as regularly as independent sensors and it may require precise timing synchronization between sensing and reconfiguration or communication. An alternative solution could be an independent sensing mechanism, working concurrently in conjunction with the transceiver. In [24] Hur *et al.* proposed a dual-stage spectrum sensing - a coarse and a fine spectrum sensing, with sensing block independent of transceiver. The coarse sensing adopts a wavelet transform to provide multi-resolution sensing over wide frequency range while the fine

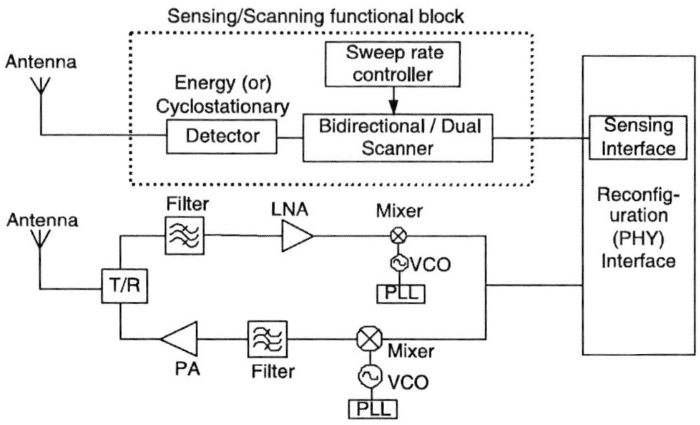

Figure 3. Transceiver architecture with external sensing and scanning mechanism.

sensing adopts Temporal Signature Detection (TSD) to exploit the periodicity of signals.

This argument could be further extended with how autonomous the receiver node should be and the type of network it is operating in. In a centralized network, sensing and information based decision making are left to the base station, so conventional receiver equipment is less intelligent as far as autonomous sensing is concerned. On the other hand, receiver equipment could conserve battery power in centralized networks. For a distributed cognitive radio network, the receiver node is expected to be autonomous, and intelligent in analyzing RF environment and decision making.

Figure 3 illustrates a simple transceiver architecture with an independent sensing and scanning block. In this architecture, the wideband signal is received through the RF front-end, sampled by the high speed analog-to-digital (A/D) converter. The RF front-end should have the capability to detect a weak signal in a large dynamic range. However, this capability requires multi-GHz speed A/D converter with high resolution, which might be infeasible [25]. The implementation of RF wideband front-end, multi-GHz A/D converter and phase locked loop (PLL) settling time during reconfigurability are some of the critical issues in cognitive radio networks [26].

3.1. Enabling Hardware Prototypes and Architectural Specifications

In [27] Yuan *et al.* presented KNOWS, a prototype cognitive wireless networking system. KNOWS is a hardware-software platform that includes a scanner radio for the reconfigurable radio, control channel based MAC protocol (CMAC) and a spectrum allocation scheme.

The requirements on the RF front-end components to learn and adapt are evident. To ease the requirements and improve the receiver performance, an image reject receiver, specifically a Weaver architecture [28], is considered. This architecture distinguishes the desired signal and the image signal by their differing phase. Thus it is possible to cancel the image signal without passing the desired signal through channel select filters. The problem with the Weaver architecture is that its image rejection capability is limited by the gain and phase mismatches in the I/Q chain. The authors of [29] attempt to provide additional image rejection by modifying the Weaver architecture to provide quadrature outputs. In [30] Luu and Daneshrad proposed a Weaver architecture with sensing capabilities to relax requirements on RF components. The learning and adapting capabilities are enabled by means of a flexible receiver architecture employing variable local oscillators (LOs) and quadrature Weaver receiver.

4. Dynamic Spectrum Management

The under-utilization of spectrum sparked the whole new paradigm of dynamic spectrum management in opposite to current static spectrum management. The term "dynamic spectrum management" has broad connotations that encompass various approaches to spectrum reform. One approach is Dynamic Spectrum Allocation and the other is an open sharing model. This model maintains the basic structure of the current spectrum regulation policy but introduces flexibility to improve spectrum efficiency. It aims to improve spectrum efficiency through dynamic spectrum assignment or access by exploiting the spatial and temporal traffic variations to different services.

Dynamic spectrum management has been analyzed through different methods based on the network types and cooperation involved. In this section, cognitive radio channel access is analyzed for centralized networks and distributed networks. In centralized networks the base station (BS) or the access point (AP) controls the allocation of channels to the cognitive radios in the network.

In distributed networks, the autonomous cognitive radio selects the channel and begins accessing it. The ability of a cognitive radio to adaptively switch between channels: *spectrum mobility* - offers tremendous scope to optimize performance. In the following, an overview of different spectrum allocation schemes for both centralized and distributed networks are discussed.

4.1. Spectrum Allocation in Centralized Networks

Spectrum allocation in centralized cognitive radio networks is relatively simple, mainly due to the decades of research in the adaptive wireless systems. Spectrum allocation is coordinated by the network moderator, which deploys mechanism-based resource management to determine the amount of transmission time to be allocated to various users on different channels such that the global performance metrics are optimized. The important question is the efficiency of such allocation and spectrum pricing (bargaining and auctioning) issues. The objective of the dynamic spectrum allocation mechanism is to maximize the utilization and this is possible only by including the spectrum agile operating characteristics of cognitive radio into the optimization strategy.

Spectrum allocation in centralized cognitive radio networks should take into account factors such as interference to primary users, fairness among sharing secondary users and overall spectrum utilization. The optimal spectrum allocation can be analyzed through graph theory and optimization techniques. In centralized allocation, the CRs dynamically compete and pay for the resources. Some of the research articles on pricing, bargaining and auctioning to maximize the revenue are summarized below.

4.1.1. Resource Allocation as Utility Maximization Problem

Many of the resource allocation problems for cognitive radios can be formulated as constrained optimization problems and solved from the network point of view or individual cognitive radio point of view. The centralized optimization can serve as the performance upper bound for the other approaches and can also provide insights for the design of other schemes. A typical example of such a constrained optimization problem is given below:

$$\text{Maximize} \sum_i U_i(\gamma_i)$$

$$\text{Subject to} \sum p_i \leq \bar{P}$$

In this example maximizing sum-utility is the objective function where utility is a function of SINR (γ) with the power constraint $\bar{P}$ in the shared spectrum. The solving methods such as Lagrangian methods, mathematical programming (e.g., linear programming, nonlinear programming and dynamic programming) can be used to obtain the solution for the formulated problem depending on the nature of objective function or constraints. If the decision variable is discrete, the problem is solved using integer and combinatorial techniques.

A discrete resource utility function and maximization of the aggregate utility by dynamically assigning network resources was studied in [31]. However, this centralized allocation method passively adjusts the resources based on previous observations and the dynamics of user behavior is not taken into account. The dynamic nature of the channel and user movement is taken into account in the utility function in [32]. The spectrum allocation in a multi-user scenario is formulated as a sum-utility maximization problem. The users' private information vector is updated in the central spectrum moderator (CSM) which then allocates resources in a socially-optimal way.

4.1.2. Resource Allocation as Graph Theory Problem

An alternative to utility maximization technique is the graph coloring method derived from graph theory. In [34] the channel allocation problem was reduced to a variant of a graph coloring problem. This paper considers a complex utility function which includes reward, conflict free assignment and a radio interference limit. This complex optimization problem is reduced to a variant of a graph coloring problem by mapping spectrum channels into colors, and assigning them to the users (vertices of graph). For example, the tripod of users whose interference range partly overlap with each other is drawn as shown in Fig 4. The bidirectional graph $G = (V, L, E)$ where V is set of vertices or users, L is available spectrum or color list, and E is the unidirected lines or interference between two vertices. In Fig 4, the channel options (A, B) are color lists for the three subnets which form the vertices. The spectrum allocation problem is equivalent to coloring each vertex using a number of colors from its color list to maximize system utility and this problem is called *color-sensitive graph coloring* (CSGC).

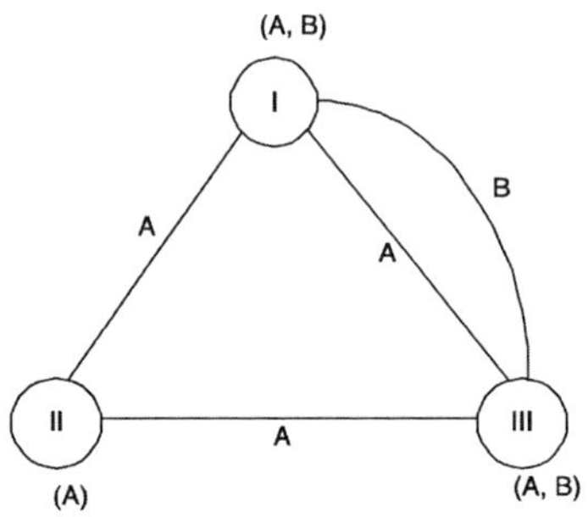

Figure 4. An example graph showing three subnets (I, II and III) with channel options A and B.

4.1.3. Resource Allocation as Pricing - Brokering problem (Spectrum Broker)

Another important approach to centralized allocation is pricing based. The centralized server could allow the users to compete and pay for the resources they wish to use. In this type of modeling, the cost per unit bandwidth (and hence the overall revenue) is maximized by pricing and auctioning rather than using the throughput/SINR values as objective functions.

In this model, a centralized operator (base station or access point) competes for the spectrum and the customers. Through demand responsive pricing, the operators compete for the resources while trying to maximize the profits. In [35], the authors have considered a system where the operator competes for the spectrum and customers. The spectrum Policy Server (SPS) performs the centralized allocation of the spectrum in a specified geographical locality in addition to maintaining the policies. The competition between the operators is formulated as a non-cooperative game and an iterative bidding scheme that results in a Nash equilibrium is proposed. In [36] a similar architecture called DIMSUMnet (Dynamic Intelligent Management of Spectrum for Ubiquitous Mobile-access network) was proposed for centralized cognitive radio networks. In DIMSUMnet the four elements, base station, clients, Radio Access Networks Manager (RANMAN) and a per-domain spectrum broker controls the spectrum allocation procedures. It uses a centralized regional level brokering mechanism where the spectrum leases the spectrum and RANMAN controls the leases for several base stations.

4.2. Spectrum Access in Distributed Networks

Channel access in distributed type cognitive radio networks is also known as Dynamic Spectrum Access (DSA). DSA is a new paradigm whereby a cognitive radio device opportunistically accesses the unutilized or under-utilized spectrum bands.

In DSA the autonomous cognitive radio node assesses the suitability of a channel, begins access and adjusts the parameters (transmission power, modulation, codebook etc.,) and policies adhering to the operating band. Most importantly the property of *spectrum agility* or *spectrum mobility* - the ability of a cognitive radio to adaptively switch between channels - offers tremendous scope to maximize performance. The key research questions are how efficiently the spectrum occupancy could be improved, the nature of coordination between the cognitive radios, the cooperative or non-cooperative communication in the network and so on.

The agility in accessing different bands with changing parameters needs better modeling techniques. The power variation in the shared channel or channel selection are analyzed using game theoretical models to study if any equilibrium solution to the formulated problem exists. The dynamic spectrum access and learning can also be analyzed based on Markov decision processes. There are reinforcement learning techniques (No-regret or Q learning) which include the dynamics as part of learning and reach the equilibrium solutions.

4.2.1. Dynamic Spectrum Access as a Game

As the distributed cognitive radios are essentially autonomous agents that learn their environment and adaptively optimize their performance by modifying the transmission parameters, their strategic decision and interactions can be modeled using a game theoretic framework. Game theory is a mathematical framework that provides a natural platform to study the effects of players' decision strategies and equilibrium solutions in a competitive environment with limited resource constraints. In this modeling of dynamic spectrum access as a game, each cognitive radio is self-interested and trying to optimize its utility function, where the utility function represents the cognitive user's performance and controls the outcomes of the game. In this model, the autonomous cognitive radios are modeled as players, their actions of selecting a parameter as strategies, and the corresponding performance changes as payoffs. Then dynamic spectrum access is modeled as a noncooperative game $\Gamma = (N, \{S\}, \{U\})$

where N is the number of players, S is the set of strategy profiles for each user and U is the utility payoff of each user.

In analyzing the game, as the players make the decisions independently the outcome of the game is influenced by other players' decisions. The utility function U_i is a function of s_i, the strategy of the user i and s_{-i}, the strategy of all other players. If s_i^* is a Nash equilibrium strategy, then

$$U(s_i^*, s_{-i}^*) \geq U(s_i, s_{-i}^*) \tag{5}$$

In the game theoretical model, the optimizing parameter, constraints and the definition of utility function defines the nature of resulting equilibrium solutions. The are several possible choices of utility functions.

- Power control game: The power control scheme for wireless networks has been extensively studied. In cognitive radio networks the transmission power under the constraints of interference or total sum power in the shared band is considered and the equilibrium solutions are obtained. A typical power control objective function is given by

$$\max_{\{p:p_i \in P_i \forall i\}} \sum_{i=1}^{N} u_i(\gamma_i(p)) \tag{6}$$

  where the power constraints is given by $p_i \in P_i = [P_i^{min}, P_i^{max}]$ and the utility is a function of SINR (γ). In [42] R. Etkin *et al.* studied the efficiency and fairness of self-enforcing spectrum sharing rules. The outcome of this noncooperative game was analyzed for one shot game and then for repeated game to emulate long period of interactions in the network. In this repeated game, the fair and efficient operating points can be enforced through punishment strategies. The results quantify the best achievable performance in a noncooperative network. A distributed power control algorithm for multichannel wireless networks is presented in [43]. In these algorithms, users announce price to reflect their sensitivities to interference levels, and then adjust their power to maximize their surplus. The convergence of the algorithm and ability to reach optimal power allocations were characterized for certain cases.

- Channel selection game: In many cases the dynamic range $[P_i^{min}, P_i^{max}]$ is limited. Also, recent developments in shared spectrum regulations suggest that the wireless nodes would have stringent power constraints. A wireless node transmitting at its highest possible power to counterbalance the interference is not always a pragmatic solution. The other method is to select the "best channel" and the channel selection is modeled as a game. The choices of transmitting channels is modeled as strategy profile and the equilibrium solutions are obtained in [44].

4.2.2. Learning Algorithms

In a distributed cognitive radio network, the game solutions can be achieved through adaptive learning algorithms. The learning algorithms for distributed dynamic spectrum access can be developed based on finite-state Markov Decision Process (MDP) and reinforcement learning such as Q-learning. In this context, a class of no-regret learning algorithms is attracting a great deal of attention. Regret-based learning procedures are simple to implement and their convergence properties are well understood. For this class of algorithms, the probability for a player i to take action $\mathbf{r}_i$ is a linear function of the *regret* which can be interpreted as the average payoff that user i would have obtained if it had played $\mathbf{r}_i^{'}$ instead of $\mathbf{r}_i$. There are several regret minimization algorithms to achieve game theoretic equilibria [45]. A more general class of no-regret learning algorithm is called Φ-no-regret learning algorithm relating to Φ-equilibria. In [44] Nie *et al.* applied this Φ-no-regret learning algorithm to a dynamic spectrum access protocol modeled as a game. By following the proposed learning adaptation process, users learn to choose the frequency channels which maximize their rewards through repeatedly playing the game.

5. Conclusion

This chapter has provided an overview of cognitive radio wireless networks and some important technical issues in realizing the technology. The research theme of cognitive radio is multidisciplinary in nature with the technical areas spanning from the physical layer, signal processing, information theory to medium access, coexistence and RF front-end enabling components. The conventional sensing methods were introduced and their advantages and disadvantages have been discussed. A number of cooperative and collaborative sensing techniques were discussed where cognitive radios perform local spectrum sensing independently and fuse their local decisions to improve the probability of detection even in the fading environments. The receiver architectures in sensing and scanning wide band spectrum and the effects on RF front-end components were studied. This is one of the key research area which would fully enable/realize the cognitive radio network. Dynamic spectrum allocation and access has been an active area of research since the adaptive access and effectiveness of the spectrum utilization depends on the mechanism. The various methods of analyzing the spectrum access techniques were studied for both centralized and distributed cognitive radio networks.

The visionary cognitive area networks is an evolving research topic and only an overview of selected topics were covered in this chapter. Adaptive medium access protocols, coexistence and coordination among the cognitive radios and link maintenance and seamless channel access are some other active areas of research. Interested readers are referred to [17] [18] [26].

Acknowledgment

The authors would like to thank Toshiba Research Europe Limited and EPSRC for funding this work through the Dorothy Hodgkin Postgraduate Award.

References

[1] Ofcom, Spectrum Framework Review, Sep 2005. Available online at http://www.ofcom.org.uk/consult/condocs/sfr/sfr2/

[2] McHenry, M. "Report on Spectrum Occupancy Measurement," Shared Spectrum Company, Aug 2005. Available online at http://www.sharedspectrum.com/?section=nsf_measurements

[3] Federal Communications Commission, "Spectrum Policy Task Force," *ET Docket* No. 02-135, Nov 2002.

[4] *Facilitating Opportunities for Flexible, Efficient, and Reliable Spectrum Use Employing Cognitive Radio Technologies,* Notice of Proposed Rule Making 03-322A1, FCC, Dec 2003.

[5] Tandra, R. and Sahai, A. "Fundamental limits on detection in low SNR under noise uncertainty," *Proceedings of WirelessCom,* Jun, 2005.

[6] Sahai, A., Hoven, N., Tandra, R. "Some Fundamental Limits on Cognitive Radio," *42nd Annual Allerton Conference on Communication, Control and Computing*, Oct 2004.

[7] J. Proakis, *Digital Communications,* 4th Ed, McGraw-Hill, 2001.

[8] Oner, M. and Jondral, F. "Cyclostationary based air interface recognition for software radio systems," *IEEE Radio and Wireless Conference,* Sept 2004

[9] Tian, Z. and Giannakis, G. "A Wavelet Approach to Wideband Spectrum Sensing for Cognitive Radios," *Third International Conference on Cognitive Radio Oriented Wireless Networks and Communications (CROWN-COM),* May, 2008.

[10] Ghasemi, A. and Sousa, E. "Collaborative Spectrum Sensing for Opportunistic Access in Fading Environments," *1st IEEE Conference on Dynamic Spectrum Access Network (DySPAN),* Nov, 2005.

[11] Mishra, S., Sahai, A. and Brodersen, R. "Cooperative Sensing among Cognitive Radios," *International Conference on Communications (ICC)*, June, 2006.

[12] Ganesan, G. and Li, Y. "Cooperative Spectrum Sensing in Cognitive Radio Networks," *1st IEEE Conference on Dynamic Spectrum Access Network (DySPAN),* Nov, 2005.

[13] Ganesan, G. and Li, Y. "Agility improvement through cooperative diversity in cognitive radio," *IEEE Globecom*, Nov 2005.

[14] Ahmed, N., Hadaller, D. and Keshav, S. "GUESS: Gossiping updates for efficient spectrum sensing," *International workshop on decentralized resource sharing in mobile computing and networking*, 2006.

[15] Gandetto, M. and Regazzoni, C. "Spectrum Sensing: A distributed Approach for Cognitive Terminals," I*EEE Journal on Selected Areas in Communications (JSAC)*, April 2007.

[16] Ghasemi, A. and Sousa, E. "Capacity of Fading Channels Under Spectrum-Sharing Constrains," *International Conference on Communications (ICC)*, June, 2006.

[17] Hossain, E. and Bhargava, V. (Eds), "Cognitive Wireless Communication Networks," Springer 2007.

[18] Haykin, S. "Cognitive Radio: Brain-Empowered Wireless Communications", *IEEE Journal on Selected Areas in Communications (JSAC)*, Feb 2005.

[19] Thomson, D.J. "Spectrum Estimation and Harmonic Analysis," *Proc IEEE,* vol.20, pp.1055-1096, Sep 1982.

[20] Thomson, D.J. "Multitaper Analysis of Nonstationary and Nonlinear Time Series Data," *Nonlinear and Nonstationary Signal Processing,* W. Fitzgerald, R. Smith, A. Walden, P. Young (Eds), Cambridge Univ. Press, 2000.

[21] Harris, F. "On Detecting White Space Spectra for Spectral Scavenging in Cognitive Radios," *Wireless Personal Communications,* vol.45, pp 325-342, 2008.

[22] Rojas, M., Lagunas, M. and Perez, A. "Candidate Spectral Estimation for Cognitive Radio," *Proc. of the 11th WSEAS Intnl. Conference on Communications,* Jul 2007.

[23] Subramani, S., Armour, S., Kaleshi, D. and Fan, Z. "Spectrum Scanning and Reserve Channel Methods for Link Maintenance in Cognitive Radio Systems," *IEEE Vehicular Technology Conference (VTC),* May 2008.

[24] Hur, Y., Park, J., Woo, W. and Lee, J., Kim, K. Lee, C., Kim, H., Laskar, J. "A Cognitive Radio (CR) System Employing a Dual-Stage Spectrum Sensing Technique: A Multi-Resolution Spectrum Sensing (MRSS) and A Temporal Signature Detection (TSD) Technique," *IEEE Globecom,* Nov, 2006.

[25] Cabric, D., Brodersen, R. "Physical Layer Design Issues Unique to Cognitive Radio Systems," *IEEE International Symposium on Personal, Indoor and Mobile Radio Communications (PIMRC)*, Sep, 2005.

[26] Akyildiz, I., Lee, W., Vuran, M. and Mohanty, S. "NeXt Generation/ dynamic spectrum access/ cognitive radio wireless networks: A survey," *Elsevier Journal of Computer Networks,* Sep, 2006.

[27] Yuan, Y., Bahl, P., Chandra, P. and Chou, P. "KNOWS: Kognitiv Networking Over White Spaces," *2nd IEEE Conference on Dynamic Spectrum Access Network (DySPAN)*, April, 2007.

[28] Weaver, D. "A Third Method of Generation and Detection of Single Sideband Signals," *Proc. IRE,* vol.44, pp.1703-1705, Dec 1956.

[29] Lee, T., Samavati, H. and Rategh, H. "5-GHz CMOS Wireless LAN," *IEEE Trans. Microwave Theory Tech.*, vol.50, pp. 268-280, Jan 2002.

[30] Luu, L. and Daneshrad, B. "An Adaptive Weaver Architecture Radio With Spectrum Sensing Capabilities to Relax RF Component Requirements," *IEEE Journal on Selected Areas in Communication (JSAC),* Apr 2007.

[31] Curescu, C. and Nadim-Tehrani, S. "Time-aware utily-based resource allocation in wireless networks," *IEEE Transactions on Parallel and Distributed*, vol.16, no.7, pp.624 - 636, 2005.

[32] A. Fattahi, F. Fu, M. Schaar, F. Paganini, "Mechanism-Based Resource Allocation for Multimedia Transmission over Spectrum Agile Wireless Networks," *IEEE Journal on Selected Areas in Communications*", April 2007.

[33] Yu, W. "Competition and Cooperation in Multi-user Communication Environments," Ph.D. dissertation, Stanford University, 2002.

[34] Peng, C., Zheng, H. and Zhao, B. "Utilization and Fairness in Spectrum Assignment for Opportunistic Spectrum Access," *ACM Journal of Mobile Networks and Applications (MONET)*, Aug, 2006.

[35] Ileri, O., Samardzija, D., Sizer, T. and Mandayam, N. "Demand Responsive Pricing and Competitive Spectrum Allocation via a Spectrum Server," 1st *IEEE Conference on Dynamic Spectrum Access Network (DySPAN)*, Nov, 2005.

[36] Buddhikot, M., Kolodzy, P., Miller, S., Ryan, K. and Evans, J. "DIMSUMnet: New Directions in Wireless Networking Using Coordinated Dynamic Spectrum Access," IEEE International Symposium on a World of Wireless, *Mobile and Multimedia Networks (IEEE WoWMoM 2005)*, Jun 2005.

[37] Unlicensed Operation in the TV Broadcast Bands and Additional Spectrum for Unlicensed Devices Below 900 MHz in the 3 GHz band. *Notice of Proposed Rulemaking 04-113*, FCC, May 2004.

[38] IEEE 802.22 *Working Group on Wireless Regional Area Networks*, http://www.ieee802.org/22

[39] Codeiro, C., Challapali, K., Ghosh, M. “Cognitive PHY and MAC layers for Dynamic Spectrum Access and Sharing of TV Bands,” *1st International Workshop on Technology and Policy for Accessing Spectrum (TAPAS)*, Aug, 2006.

[40] Zhao, Q., Tong, L., Swami, A., Chen, Y. “Decentralized Cognitive MAC for Opportunistic Spectrum Access in Ad hoc Networks: A POMDP Framework,” *IEEE Journal on Selected Areas in Communication (JSAC)*, Apr, 2007.

[41] Xing, Y., Chandramouli, R., Mangold, S. and Shankar, S. “Dynamic Spectrum Access in Open Spectrum Wireless Networks,” *IEEE Journal on Selected Areas in Communication (JSAC),* Mar, 2006.

[42] Etkin, R., Parekh, A. and Tse, D. “Spectrum sharing for Unlicensed Bands,” *IEEE Journal on Selected Areas in Communication (JSAC),* Apr, 2007.

[43] Huang, J., Berry, R. and Honig, M.“Distributed Interference Compensation for Wireless Networks,” *IEEE Journal on Selected Areas in Communication (JSAC)*, May, 2006.

[44] Nie, N., Comaniciu, C. “Adaptive channel allocation spectrum etiquette for congnitive radio networks,” *1st IEEE Conference on Dynamic Spectrum Access Network (DySPAN)*, Nov, 2005.

[45] Greenwald, A. and Jafari, A. “A General Class of No-Regret Algorithms and Game-Theoretic Equilibria,” *Computational Learning Theory Conference,* Aug 2003.

In: Cognitive Radio
Editors: P.D. McGuire et al, pp. 263-264
ISBN: 978-1-60876-604-8

Chapter 9

EXPERT COMMENTARY

Shilian Zheng and Caiyi Lou
JiangNan Electronic Communication Research Institute,
Jiaxing 314033, China

Cognitive radio is regarded as a radio that can sense, aware, learn and adapt to the outside environment. Evolutionary algorithms have been shown as a promising technique for adaptation in this book. However, their applications in cognitive radios do not restrict to adaptation at all. We have now been working on applying evolutionary algorithms to solve other problems in cognitive radios.

Spectrum sensing, as a key enabling functionality in cognitive radio networks, needs to reliably detect weak primary radio signals. In order to improve the reliability of spectrum sensing, cooperative spectrum sensing has been proposed in the literature. In such scenarios, a network of cooperative cognitive radios, which experience different channel conditions from the target, would have a better chance of detecting the primary radio if they combine the sensing information jointly. As such, cooperative spectrum sensing can be formalized as an optimization problem in the fusion center which can be easily tackled by evolutionary algorithms.

Dynamic spectrum assignment is also a key enabling functionality in cognitive radio networks. After spectrum sensing, cognitive radios obtain information of spectrum holes in the network. Dynamic spectrum assignment refers to the problem of how to share these spectrum holes among cognitive radios as the wireless environment changes without causing harmful interference to one another or minimize the interference of the network. Based on centralized or distributed architecture, cooperative or non-cooperative spectrum allocation behavior, overlay or underlay spectrum access technique, lots of methods have been proposed for dynamic spectrum access, including game theory, pricing and auction mechanisms, local bargaining, and graph coloring. As spectrum allocation model can be inherently seen as an optimization problem, it can be solved by evolutionary algorithms. Our preliminary experiments have shown their effectiveness.

Many of the algorithms in cognitive radio networks require the existence of a commonly known and globally available control channel through which information is distributed among the network nodes. Without such information, the cognitive radio algorithms' performance is significantly deterred or the algorithm may not function properly at all. Thus, possessing such a common global control channel for active coordination is crucial for many applications. C. Doerr et al. have proposed to use swarm intelligence as a way to dynamically find and manage such control channels in cognitive radio networks. This problem can also be approached by evolutionary algorithms. Whether there is performance improvement needs further exploration.

It is rational to believe that evolutionary algorithms would play an important part in cognitive radios due to their intelligent nature in solving optimization problems. They may also be effective in approaching power control problem and learning aspects of cognitive radios. The current networks may also be evolved to be cognitive, which means that a cognitive network has the ability to perform its tasks in an autonomous way by using its attributes such as self-managing, self-optimizing, self-monitoring, self-repair, self-protection, self-adaptation, self-healing to adapt to changing requirements or component failures. Evolutionary algorithms may be a good solution for cognitive networking. These applications are worth further exploration.

INDEX

A

B

D

E

K

L

M

N

O

P

Q

R

S

T

U

V

W